Springer Proceedings in Mathematics & Statistics

Volume 552

AF843135

This book series features volumes composed of selected contributions from workshops and conferences in all areas of current research in mathematics and statistics, including data science, operations research and optimization. In addition to an overall evaluation of the interest, scientific quality, and timeliness of each proposal at the hands of the publisher, individual contributions are all refereed to the high quality standards of leading journals in the field. Thus, this series provides the research community with well-edited, authoritative reports on developments in the most exciting areas of mathematical and statistical research today.

Young-Heon Kim · Soumik Pal · Brendan Pass
Editors

The Mathematics of Monge-Kantorovich Optimal Transport

PIMS-IFDS-NSF Summer School, Seattle, Washington, USA, June 19–July 1, 2022

Pacific Institute for the Mathematical Sciences

Springer

Editors
Young-Heon Kim
Department of Mathematics
University of British Columbia
Vancouver, BC, Canada

Soumik Pal
Department of Mathematics
University of Washington
Seattle, WA, USA

Brendan Pass
Departments of Mathematical
and Statistical Sciences
University of Alberta
Edmonton, AB, Canada

ISSN 2194-1009 ISSN 2194-1017 (electronic)
Springer Proceedings in Mathematics & Statistics
ISBN 978-3-032-23145-1 ISBN 978-3-032-23146-8 (eBook)
https://doi.org/10.1007/978-3-032-23146-8

© The Editor(s) (if applicable) and The Author(s), under exclusive license to Springer Nature Switzerland AG 2026

This work is subject to copyright. All rights are solely and exclusively licensed by the Publisher, whether the whole or part of the material is concerned, specifically the rights of translation, reprinting, reuse of illustrations, recitation, broadcasting, reproduction on microfilms or in any other physical way, and transmission or information storage and retrieval, electronic adaptation, computer software, or by similar or dissimilar methodology now known or hereafter developed.
The use of general descriptive names, registered names, trademarks, service marks, etc. in this publication does not imply, even in the absence of a specific statement, that such names are exempt from the relevant protective laws and regulations and therefore free for general use.
The publisher, the authors and the editors are safe to assume that the advice and information in this book are believed to be true and accurate at the date of publication. Neither the publisher nor the authors or the editors give a warranty, expressed or implied, with respect to the material contained herein or for any errors or omissions that may have been made. The publisher remains neutral with regard to jurisdictional claims in published maps and institutional affiliations.

This Springer imprint is published by the registered company Springer Nature Switzerland AG
The registered company address is: Gewerbestrasse 11, 6330 Cham, Switzerland

If disposing of this product, please recycle the paper.

Preface

This proceedings gathers written notes from three of the lecture series offered as part of the PIMS—IFDS—NSF Summer School on Optimal Transport, at the University of Washington in June of 2022. This summer school was the first large event organized by the Kantorovich Initiative, a then nascent research organization linking several universities in the Pacific Northwest, focusing on the mathematics of the Monge–Kantorovich optimal transport problem and its many applications. Organized at the tail end of the COVID pandemic, this successful summer school exposed its 150+ participants to the manifold research opportunities arising from the area of mathematical theory of Optimal Transport and its applications in biology, economics, and machine learning. Three of our lecturers, leading experts in optimal transport and its connections with various other areas, generously agreed to write notes based on their lectures (with co-authors in two cases) and these form the content of this volume.

The breadth of topics covered in this volume is striking, and reflective of the remarkable breadth of modern optimal transport research. Felix Otto and Lukas Koch's contribution, based on Otto's mini-course, describes a variational approach to the regularity theory of optimal transport. Alfred Galichon and Antoine Jacquet's notes, based on Galichon's mini-course, develop the connections between optimal transport and matching models in economics, while Geoff Schiebinger's notes describe his mini-course on applications of optimal transport in developmental biology. We strongly believe that these beautiful mini-courses and the corresponding lecture notes, will have a significant impact on the next generation of optimal transport researchers.

We note that there were also other fascinating mini-courses at the summer school, though not included in this volume: "Recent Results in Interface Motions in the Framework of Optimal Transport" by Inwon C. Kim, "Recent Advances in Dynamical Optimal Transport" by Jan Maas, and "Optimal Transport for Machine Learning" by Gabriel Peyré.

We would like to express our deep gratitude to the leadership and staff at the Pacific Institute of Mathematical Sciences (PIMS). Without their generous support and guidance there would be no Kantorovich Institute and this summer school could not have taken place. In addition to PIMS, we would also like to acknowledge financial support from the National Science Foundation (NSF), the Institute for the Foundations of Data Science (IFDS) at the University of Washington, the Department of Mathematical and Statistical Sciences at the University of Alberta, the Department of Mathematics at the University of British Columbia and the Departments of Applied Mathematics, Economics, Mathematics, and Statistics at the University of Washington.

Vancouver, Canada — Young-Heon Kim
Seattle, USA — Soumik Pal
Edmonton, Canada — Brendan Pass

Contents

Substitutability, Equilibrium Transport, and Matching Models

Alfred Galichon and Antoine Jacquet

Abstract This chapter surveys how *substitutability* organizes computation in a large class of models encompassing optimal transport and matching. We review Z-, M_0-, and M-functions and inverse-isotonicity results, showing when Jacobi's algorithm (including Sinkhorn/IPFP) converge. For transferable and imperfectly transferable settings, a distance-to-frontier formulation recasts equilibria as M- or M_0-function fixed points, clarifying existence and uniqueness. For non-transferable utility, we reinterpret Gale–Shapley and Adachi, expose the lattice of stable matchings, and present DALM for large markets. Toy applications–hedonic surge pricing and rent-controlled housing–ground the methods. Overall, substitutability provides a powerful alternative to convexity for theory and computation.

1 Introduction

This chapter is about *substitutability*. Substitutability is a fundamental property of optimal transport models, although it is less recognized than convexity.

Consider the problem of computing an economic equilibrium

$$Q(p) = 0 \tag{1}$$

where p is a vector of prices for n goods, and $Q : \mathbb{R}^n \to \mathbb{R}^n$ is an excess supply function (supply minus demand for each of the n goods).

Broadly speaking, there are only two categories of economic models that can be conveniently computed. The first category includes the models that rely on convexity, i.e. models where $Q = \nabla U(p)$ with U a convex function. These models can be

A. Galichon (✉)
New York University, New York, USA
e-mail: alfred.galichon@nyu.edu

A. Galichon · A. Jacquet
Sciences Po, Paris, France
e-mail: antoine.jacquet@sciencespo.fr

© The Author(s), under exclusive license to Springer Nature Switzerland AG 2026

Y. Kim et al. (eds.), *The Mathematics of Monge-Kantorovich Optimal Transport*, Springer Proceedings in Mathematics & Statistics 552,
https://doi.org/10.1007/978-3-032-23146-8_1

reformulated as an optimization problem (minimization of U) and *descent methods*, such as the standard gradient descent, will work:

$$p_z^{t+1} = p_z^t - \varepsilon\, Q_z\left(p^t\right), \quad \varepsilon > 0 \text{ small.} \tag{2}$$

The second category comprises the models that rely on substitutability – essentially the idea that Q_z is increasing in p_z and decreasing in p_x for $x \neq z$. In this case, *coordinate update methods*, such as nonlinear Jacobi, where we set the price of good z to clear the market for good z, will work:

$$Q_z\left(p_z^{t+1}, p_{-z}^t\right) = 0. \tag{3}$$

By an extraordinary coincidence, optimal transport inherits both structures. For this reason, one can compute the OT problem and its regularizations both by descent methods and by coordinate update methods (which is called Sinkhorn's algorithm).

Some matching models can be computed using optimal transport. These are called transferable utility models, and they assume that everyone's valuations are expressed in the same monetary unit. In that case, equilibrium in matching problems is the solution to an optimization problem. However, in many other cases this assumption cannot be made. In labor economics, for instance, one dollar for the firm does not have the same value as one dollar for the employee (pre-tax or post-tax; decreasing marginal utility, etc). Similarly, in family economics partners may transfer utility by allocating public and private expenditures that can be inefficient. Lastly, in school choice problems utility is cardinal and cannot be transferred.

We will see mathematically that these models *cannot* be recast as optimization problems, and so the convex optimization structure is lost. They can, however, be reformulated as models with substitutability, from which we will be able to deduce a lot of structure and computational methods.

The chapter is organized as follows. We will start in Sect. 2 by introducing Jacobi's algorithm and some of its basic properties. We will proceed with a rather detailed study of substitutability and its mathematical counterparts, *Z- and M-functions*. There was a vibrant literature on this topic in the 1970s (Birkhoff, Rheinboldt, Ortega, More, Porsching, and some others; see in particular Ortega and Rheinboldt's 1970 book [1]), but this literature has somehow fallen out of attention. We will unearth some of the main results from that literature regarding the convergence of Jacobi's algorithm, and provide some new ones. The section ends with a toy hedonic model as an illustration.

In Sect. 3, we will appeal to the machinery developed in Sect. 2 to study models of matching with transfers, departing from the assumption of perfectly transferable utility. The section will deal with the regularized and unregularized cases, and it will show the convergence of IPFP, and discuss the existence and uniqueness of an equilibrium.

In Sect. 4, we will revisit Gale and Shapley's famous theory of "stable marriages," which is the theory of matchings without transfers, and we will reinterpret the well-known deferred acceptance algorithm as a damped version of Jacobi's algorithm. We also present Adachi's formulation of the stable matching problem, which leads to the more recent notion of equilibrium matching.

We hope this chapter will provide a solid mathematical introduction to matching models in economics and econometrics.

2 M-Functions and Jacobi's Algorithm

In this section we introduce the mathematical notions related to models with substitutability, notably Z- and M-functions, as well as Jacobi's algorithm for solving such models.

2.1 Jacobi's Algorithm

Put simply, our goal is to solve a system of nonlinear equations

$$Q(p) = 0 \tag{4}$$

where $Q : \mathbb{R}^Z \to \mathbb{R}^Z$ and $|Z| = n$. We shall assume throughout:

Assumption 2.1 (Continuity) Q is continuous.

In economic terms, Q can be interpreted as an excess supply function. Suppose the demand for good z is a function $D_z(p)$ which depends on all prices, and similarly the supply for good z is a function $S_z(p)$. Then

$$Q_z(p) = S_z(p) - D_z(p) \tag{5}$$

is the excess supply for good z. Solving $Q_z(p) = 0$ for all z is thus akin to finding a vector of prices p which clears the markets for all n goods simultaneously.

Jacobi's algorithm is a method to solve such a system. It consists of taking an initial guess p^0 and defining a sequence (p^t) iteratively according to

$$p_z^{t+1} : \quad Q_z\left(p_z^{t+1}, p_{-z}^t\right) = 0. \tag{6}$$

Here p_{-z} denotes the vector p without its zth entry, and (π, p_{-z}) the vector p with its zth entry replaced by π. The mapping $T : p^t \mapsto p^{t+1}$ is called the *coordinate update* or *Jacobi update* operator. To ensure that this operator is well-defined, we shall assume throughout:

Assumption 2.2 (Responsiveness) For all z,

$$\inf_{\pi} Q_z(\pi, p_{-z}) < 0 < \sup_{\pi} Q_z(\pi, p_{-z}), \quad \forall p_{-z}. \tag{7}$$

We also remove any ambiguity in choosing p_z^{t+1} by setting precisely

$$p_z^{t+1} = \min\left\{\pi : Q_z(\pi, p_{-z}^t) = 0\right\}. \tag{8}$$

Jacobi's algorithm has an intuitive economic interpretation. Let's assume that each good z has an auctioneer in charge of determining its "right" price p_z. At each period, the auctioneer for good z sets p_z in order to clear the corresponding market, but using the prices from the previous period (it is as if each auctioneer did not anticipate that other prices were being updated simultaneously). Because of this, once the simultaneous price updates are taken into account, the price chosen by the auctioneer is in fact typically not market clearing. This process is then re-iterated until the prices converge.

As an illustration, consider the regularized optimal transport problem

$$\begin{aligned}\max_{\mu\geq 0} &\sum_{xy\in X\times Y} \mu_{xy}\Phi_{xy} - \sum_{xy\in X\times Y} \mu_{xy}\log\mu_{xy} \\ \text{s.t.} &\sum_{y\in Y}\mu_{xy} = n_x, \quad \sum_{x\in X}\mu_{xy} = m_y,\end{aligned}$$

which has solutions of the form: $\mu_{xy} = \exp\left(\Phi_{xy} - u_x - v_y\right)$. A common interpretation is that of a two-sided matching market between workers x and firms y who respectively ask for utility u_x and v_y. We will come back to this problem and its interpretation in Sect. 3.

A standard procedure to solve this type of problem is Sinkhorn's algorithm, which is nothing else than Jacobi's algorithm adapted to this special case.

Sinkhorn's algorithm

Take an initial guess for u^0 and v^0, and set for $t \geq 0$:

$$\begin{cases} u_x^{t+1} : n_x = \sum_y \exp\left(\Phi_{xy} - u_x^{t+1} - v_y^t\right) \\ v_y^{t+1} : m_y = \sum_x \exp\left(\Phi_{xy} - u_x^t - v_y^{t+1}\right)\end{cases}$$

until approximate convergence is reached. (Notice that here, u^{2t} and v^{2t+1} depend exclusively on the starting point u^0, while u^{2t+1} and v^{2t} depend exclusively on the starting point v^0.)

It is easy to see that by defining $p_x = -u_x$, $p_y = v_y$, as well as

$$\begin{aligned} Q_x(p) &= \sum_y \exp\left(\Phi_{xy} + p_x - p_y\right) - n_x \\ Q_y(p) &= -\sum_x \exp\left(\Phi_{xy} + p_x - p_y\right) + m_y,\end{aligned}$$

we can rewrite Sinkhorn's algorithm exactly as (6).

Note that Q_x is an antitone function with respect to p_{-x} (i.e. nonincreasing with respect to the componentwise order), and similarly Q_y is an antitone function with respect to p_{-y}. We will see below how this property plays an important role in Jacobi's algorithm.

One important question remains: does the Jacobi sequence (p^t) converge? Although this is not the case in general, this sequence does converge when the model exhibits several features related to substitutability. The rest of this section will be dedicated to defining exactly what we mean by substitutability, and to studying this category of models – and notably the behavior of the Jacobi sequence.

2.2 Nondecreasing Jacobi Sequence

The convergence of the Jacobi sequence will rely on the notion of subsolution, that we define as follows:

Definition 2.3 *We say that p is a subsolution if $Q(p) \leq 0$, and that p is a supersolution if $Q(p) \geq 0$.*

Broadly speaking, we expect to interpret a subsolution as "prices are too low," and a supersolution as "prices are too high." We shall develop this intuition below.

Let p^t be a subsolution and assume p^{t+1} exists, so that

$$Q_z\left(p_z^t, p_{-z}^t\right) \leq 0, \; Q_z\left(p_z^{t+1}, p_{-z}^t\right) = 0. \tag{9}$$

Then we can guarantee that $p^t \leq p^{t+1}$ using:

Assumption 2.4 (Diagonal isotonicity) $Q_z(p)$ is nondecreasing in p_z.

We state this result formally:

Proposition 2.5 *Assume Q is diagonal isotone. If p^t is a subsolution, then $p^t \leq p^{t+1}$.*

Proof. $Q_z\left(p_z^t, p_{-z}^t\right) \leq 0 = Q_z\left(p_z^{t+1}, p_{-z}^t\right)$ and Q_z is increasing in p_z. □

Next, we can guarantee that p^{t+1} is still a subsolution with:

Assumption 2.6 (Z-function) $Q_z(p_z, p_{-z})$ is antitone with respect to p_{-z}, i.e. $p_{-z} \leq p'_{-z}$ implies $Q_z(p_z, p_{-z}) \geq Q_z\left(p_z, p'_{-z}\right)$.

In economic terms, Q being a Z-function means that when the price of any other good increases, the excess supply for good z should decrease. This can be explained by a standard substitution pattern: if all prices but p_z increase, then producers should supply less good z, while consumers should demand more good z. For this reason, we also say that Q has the *substitutes property*.

With the Z-function assumption we have:

Proposition 2.7 *Assume Q is a diagonal isotone Z-function. If p^t is a subsolution, then p^{t+1} is also a subsolution.*

Proof. We have $p^t \leq p^{t+1}$ by diagonal isotonicity, so in particular $p^t_{-z} \leq p^{t+1}_{-z}$ for any z. Hence, since Q_z is antitone with respect to p_{-z}, $Q_z\left(p^{t+1}_z, p^{t+1}_{-z}\right) \leq Q_z\left(p^{t+1}_z, p^t_{-z}\right) = 0$. □

The immediate consequence of Propositions 2.5 and 2.7 is that, with Q satisfying the assumptions above, any Jacobi sequence starting from a subsolution will be nondecreasing. Yet, as the following example shows, this is still not sufficient to ensure convergence.

Example 2.8 *Consider a linear $Q(p)$ given by*

$$Q(p) = \begin{pmatrix} 1 & -2 \\ -2 & 1 \end{pmatrix} p.$$

The Jacobi sequence starting from p^0 can be written explicitly as

$$\begin{cases} p^t_1 = 2^t\, p^0_2 \\ p^t_2 = 2^t\, p^0_1. \end{cases}$$

Even though $p^0 = (1, 1)$ is a subsolution, and Q is a continuous diagonal isotone Z-function, the associated Jacobi sequence does not converge to the unique solution $(0, 0)$. In fact, it is clear from the expressions of p^t_1 and p^t_2 that any Jacobi sequence starting from a vector other than $(0, 0)$ will diverge.

2.3 M- and M$_0$-Functions

Example 2.8 shows what went wrong: the function Q allowed for an inversion by allowing the subsolution $(1, 1)$ to be above the solution $(0, 0)$. Recall that we expect subsolution to mean "prices are too low" and supersolution "prices are too high," and therefore we wish to rule out such inversions. We do so using the following notions, defined in [2]:

Definition 2.9 *Consider $Q : \mathbb{R}^n \to \mathbb{R}^n$. We say that:*

(i) Q is strongly nonreversing if

$$\begin{cases} p \geq p' \\ Q(p) \leq Q(p') \end{cases} \text{implies } p = p'. \tag{10}$$

(ii) Q is weakly nonreversing if

$$\begin{cases} p \geq p' \\ Q(p) \leq Q(p') \end{cases} \text{implies } Q(p) = Q(p'). \tag{11}$$

Economically speaking, Q being nonreversing means that when prices increase, the excess supply cannot decrease for all goods. In other words, it ensures that the price effect remains stronger than the substitution effect.

Example 2.10 *Consider $Q(p) = Qp$ with Q an $n \times n$ matrix. If Q is weakly (column) diagonally dominant (i.e. $1^\top Q \geq 0$), then Q is weakly nonreversing. If it is strictly (column) diagonally dominant (i.e. if $1^\top Q > 0$), then Q is strongly nonreversing.*

Example 2.11 *Consider Q such that $1^\top Q(p)$ is weakly isotone in p. Then it is weakly nonreversing. In particular, this is the case of $Q : \mathbb{R}^{X \cup Y} \to \mathbb{R}^{X \cup Y}$ given by*

$$\begin{cases} Q_x(p) = \sum_y \exp\left(\Phi_{xy} + p_x - p_y\right) - n_x, & x \in X, \\ Q_y(p) = -\sum_x \exp\left(\Phi_{xy} + p_x - p_y\right) + m_y, & y \in Y. \end{cases} \tag{12}$$

Example 2.12 *Consider Q such that $1^\top Q(p) \leq 1^\top Q(p')$ and $p \geq p'$ together imply $p = p'$. Then it is strongly nonreversing. In particular, this is the case of $Q : \mathbb{R}^{X \cup Y} \to \mathbb{R}^{X \cup Y}$ given by*

$$\begin{cases} Q_x(p) = \sum_y \exp\left(\Phi_{xy} + p_x - p_y\right) + \exp(p_x) - n_x, & x \in X \\ Q_y(p) = -\sum_x \exp\left(\Phi_{xy} + p_x - p_y\right) - \exp\left(-p_y\right) + m_y, & y \in Y. \end{cases}$$

Next we define M- and M_0-functions as Z-functions which are also nonreversing:

Definition 2.13 *An M-function is a Z-function which is strongly nonreversing. An M_0-function is a Z-function which is weakly nonreversing.*

Clearly, M-functions constitute a subset of M_0-functions since strong nonreversingness implies weak nonreversingness. In economic terms, M- and M_0-functions exhibit both the substitutes property (which they inherit from Z-functions), and the property of having a price effect stronger than the substitution effect.

The combination of these two properties implies that M_0-functions (and therefore M-functions also) are diagonal isotone, so that M- and M_0-functions satisfy Assumptions 2.4 and 2.6.

Proposition 2.14 *Assume Q is an M_0-function, then it is diagonal isotone.*

Proof. Let p and p' such that $p_z \leq p'_z$ and $p_{-z} = p'_{-z}$. Looking for a contradiction, suppose that $Q_z(p) > Q_z(p')$.

For any $\tilde{z} \neq z$ we have $p_{\tilde{z}} = p'_{\tilde{z}}$ and $p_{-\tilde{z}} \leq p'_{-\tilde{z}}$, so since Q is a Z-function,

$$Q_{\tilde{z}}(p) = Q_{\tilde{z}}(p_{\tilde{z}}, p_{-\tilde{z}}) \geq Q_{\tilde{z}}\left(p_{\tilde{z}}, p'_{-\tilde{z}}\right) = Q_{\tilde{z}}\left(p'\right).$$

Hence $p \leq p'$ and $Q(p) \geq Q(p')$, therefore $Q(p) = Q(p')$ by weak nonreversingness. We have a contradiction: in fact $Q_z(p) \leq Q_z(p')$. □

Galichon, Samuelson and Vernet [2] characterize M-functions using:

Theorem 2.15 (Inverse isotonicity theorems) *Consider Q a Z-function. Then:*

(i) Q is an M-function if and only if it is inverse isotone, i.e.

$$Q(p) \leq Q(p') \text{ implies } p \leq p'.$$

(ii) Q is an M_0-function if and only if the set-valued map Q^{-1} is isotone in the strong set order, i.e.

$$Q(p) \leq Q(p') \text{ implies } \begin{cases} Q(p) = Q(p \wedge p') \\ Q(p') = Q(p \vee p'). \end{cases}$$

We recall that $p \vee p'$ denotes the *join* (componentwise maximum) of two vectors p and p', while $p \wedge p'$ denotes their *meet* (componentwise minimum).

Proof. (i) Assume Q is an M-function and $Q(p) \leq Q(p')$. Then $p \vee p' \geq p'$ and

$$Q_z(p \vee p') \leq 1\{p_z \leq p'_z\} Q_z(p') + 1\{p_z > p'_z\} Q_z(p) \leq Q_z(p'),$$

therefore we conclude that $p' = p \vee p'$, and thus $p \leq p'$. Conversely, assume Q is inverse isotone. To show it is strongly nonreversing, assume $p \geq p'$ and $Q(p) \leq Q(p')$. By inverse isotonicity $p \leq p'$, hence $p = p'$.

(ii) Assume Q is an M_0-function and $Q(p) \leq Q(p')$. Then $p \vee p' \geq p'$ and

$$Q_z(p \vee p') \leq 1\{p_z \leq p'_z\} Q_z(p') + 1\{p_z > p'_z\} Q_z(p) \leq Q_z(p'),$$

therefore we conclude that $Q(p \vee p') = Q(p')$. Similarly, $p \geq p \wedge p'$ and

$$Q_z(p \wedge p') \geq 1\{p_z \leq p'_z\} Q_z(p) + 1\{p_z > p'_z\} Q_z(p') \geq Q_z(p)$$

thus $Q(p \wedge p') = Q(p)$. Conversely, assume Q^{-1} is isotone in the strong set order. To show that Q is weakly nonreversing, assume $p \geq p'$ and $Q(p) \leq Q(p')$. The former implies $p \wedge p' = p'$, and the latter $Q(p) = Q(p \wedge p')$. □

Corollary *Assume Q is an M_0-function, then the set of solutions is stable by the join and meet operations $\vee$ and $\wedge$.*

2.4 Convergence of the Jacobi Sequence

When Q is an M- or M_0-function, inversions such as the one we saw in Example 2.8 are ruled out. In this case, as long as a sub- and a supersolution exist, we can find a Jacobi sequence which converges towards a solution.

Ortega and Rheinboldt [1] prove the two following results for M-functions.

Theorem 2.16 *Assume Q is a continuous responsive M-function, and that both a sub- and a supersolution exist. Then a solution exists and it is unique, and any Jacobi sequence starting from a sub- or supersolution converges to this unique solution.*

Proof. The proof consists in considering two Jacobi sequences $(\hat{p}^t)$ and $(\check{p}^t)$ starting from respectively a sub- and a supersolution. $(\hat{p}^t)$ is an increasing sequence of subsolutions, $(\check{p}^t)$ is a decreasing sequence of supersolutions, and by inverse isotonicity, one has $\hat{p}^t \leq \check{p}^t$, and therefore they both converge to respectively p and p'. $Q(p) = Q(p') = 0$ and inverse isotonicity applied twice yields $p = p'$. □

If Q is surjective, then we need not even start from a sub- or supersolution.

Theorem 2.17 *Assume Q is also surjective. Then any Jacobi sequence converges to the unique solution.*

Proof. Let (p^t) be the Jacobi sequence starting from a vector p^0. Since Q is surjective, there exist $\hat{p}^0$ such that $Q(\hat{p}^0) = Q(p^0) \wedge 0$, and $\check{p}^0$ such that $Q(\check{p}^0) = Q(p^0) \vee 0$. Then $\hat{p}^0$ is a subsolution, $\check{p}^0$ is a supersolution, and the Jacobi sequences $(\hat{p}^t)$ and $(\check{p}^t)$ starting from these respective vectors are such that

$$\hat{p}^t \leq p^t \leq \check{p}^t$$

with $(\hat{p}^t)$ increasing and $(\check{p}^t)$ decreasing. As a result, all three sequences converge, and their limits are solutions. □

Finally, Galichon and Léger [3] obtain the following result for M_0-functions.

Theorem 2.18 *Assume Q is an M_0-function, and that both a sub- and a supersolution exist. Then a solution exists and can be obtained as the limit of the Jacobi sequence starting from the meet (or the join) of a subsolution and a supersolution.*

Proof. Let $\check{p}$ be a subsolution and $\hat{p}$ a supersolution. Then $Q(\check{p}) \leq Q(\hat{p})$, therefore the inverse isotonicity characterization of M_0-functions implies $Q(\check{p} \wedge \hat{p}) = Q(\check{p})$ and $Q(\check{p} \vee \hat{p}) = Q(\hat{p})$. Hence $\check{p} \wedge \hat{p}$ is a subsolution and $\check{p} \vee \hat{p}$ is a supersolution. Consider $(\underline{p}^t)$ and $(\bar{p}^t)$ the Jacobi sequences starting respectively from $\underline{p}^0 = \check{p} \wedge \hat{p}$ and $\bar{p}^0 = \check{p} \vee \hat{p}$, so that $\underline{p}^0 \leq \bar{p}^0$. We show by induction that $\underline{p}^t \leq \bar{p}^t$. Looking for a contradiction, assume that $\underline{p}_z^{t+1} > \bar{p}_z^{t+1}$ for some z. Then

$$0 = Q_z(\underline{p}_z^{t+1}, \underline{p}_{-z}^t) \geq Q_z(\bar{p}_z^{t+1}, \underline{p}_{-z}^t) \geq Q_z(\bar{p}_z^{t+1}, \bar{p}_{-z}^t) = 0$$

by diagonal isotonicity and the Z-function property. Hence $Q_z(\underline{p}_z^{t+1}, \underline{p}_{-z}^t) = Q_z(\bar{p}_z^{t+1}, \underline{p}_{-z}^t) = 0$ but $\underline{p}_z^{t+1} > \bar{p}_z^{t+1}$, which contradicts the definition (8) of $\underline{p}_z^{t+1}$ as the smallest solution to $Q_z(\pi, \underline{p}_{-z}^t) = 0$. Thus $(\underline{p}^t)$ is increasing, $(\bar{p}^t)$ is decreasing, and $\underline{p}^t \leq \bar{p}^t$, therefore both sequences are bounded and have a limit. Hence a solution exists. □

We now consider some examples around *constant aggregates*, i.e. the property that $1^{\top} Q(p) = \sum_z Q_z(p)$ is constant for all p.

Proposition 2.19 *Assume Q is a continuous Z-function with $1^{\top} Q(p) = 0$, and that there is a $0 \in Z$ such that the restriction and corestriction of Q to $\mathbb{R}^{Z\setminus\{0\}}$ with the normalization $p_0 = \pi$, denoted $Q_{-0} : \mathbb{R}^{Z\setminus\{0\}} \to \mathbb{R}^{Z\setminus\{0\}}$, is a surjective M-function. Then, denoting $p(\pi)$ the unique solution to $Q_{-0}(p) = 0$, one has that $\pi \leq \pi'$ implies $p(\pi) \leq p(\pi')$.*

Example 2.20 *Consider the case of regularized optimal transport:*

$$\begin{cases} Q_x(p) = \sum_y \exp\left(\Phi_{xy} + p_x - p_y\right) - n_x \\ Q_y(p) = -\sum_x \exp\left(\Phi_{xy} + p_x - p_y\right) + m_y, \end{cases}$$

so that $1^{\top} Q(p) = \sum_y m_y - \sum_x n_x$ is constant. It is easy to see that Q is a Z-function. In addition, because of constant aggregates, $Q(p) \leq Q(p')$ implies $Q(p) = Q(p')$ so Q is weakly nonreversing, thus it is an M_0-function. However, Q is not an M-function: take for instance $p = p' + 1_Z$, then we have $p \geq p'$ and $Q(p) = Q(p')$, but $p \neq p'$ so Q is not strongly nonreversing.

Example 2.21 *Consider a linear $Q(p) = (\Delta - A)p$ where Δ is a diagonal matrix with positive entries, A has a zero diagonal and positive off-diagonal entries, and*

$$1^{\top}(\Delta - A) = 0$$

so that the function Q has constant aggregates. Then the matrix $\Delta^{-1}A$ is nonnegative and irreducible (it is strongly connected), and the Perron–Frobenius theorem applies. Letting $\delta = \Delta 1$, we have $\delta \geq 0$ and $\delta^{\top} = 1^{\top}A = \delta^{\top}\Delta^{-1}A$, so δ must be the left Perron eigenvector of $\Delta^{-1}A$, and it is associated to the Perron eigenvalue 1. Thus there is also a right Perron eigenvector $v \geq 0$ such that $\Delta^{-1}Av = v$, i.e. $Q(v) = (\Delta - A)v = 0$.

2.5 Application: A Toy Hedonic Model

Consider a surge pricing problem in an Uber-like environment. We have partitioned the city in a finite number of locations (say, blocks). Let $x \in X$ denote the location of the driver, $y \in Y$ the location of the passenger, and $z \in Z$ the pickup location. Assume that for a driver at x, the cost of picking up a passenger at z is c_{xz}. If the price of the ride at z is p_z, the utility of the driver is $p_z - c_{xz} + \varepsilon_z$, where the vector (ε_z) is random. If the driver does not pick up anyone, the utility is normalized to ε_0. Assume that (ε_z) is i.i.d. Gumbel. Then the probability that a driver at x will demand a ride z is

$$\frac{\exp\left(p_z - c_{xz}\right)}{1 + \sum_{z'} \exp\left(p_{z'} - c_{xz'}\right)}. \tag{13}$$

Now assume there are n_x drivers in area x, therefore the supply for rides at z is

$$S_z(p) = \sum_{x \in X} n_x \frac{\exp(p_z - c_{xz})}{1 + \sum_{z'} \exp(p_{z'} - c_{xz'})}. \tag{14}$$

It is easy to see that $S_z(p)$ is decreasing with respect to $p_{z'}$ for $z' \neq z$.

Now let's focus on demand. This is the same as before, except for the fact that the utility of a passenger at y seeking a ride at location z is now $a_{yz} - p_z + \eta_z$, and the utility of the outside option is η_0, where (η_z) is i.i.d. Gumbel. Assuming there are m_y passengers in area y, the induced demand is

$$D_z(p) = \sum_{y \in Y} m_y \frac{\exp(a_{yz} - p_z)}{1 + \sum_{z'} \exp(a_{yz'} - p_{z'})} \tag{15}$$

and we see that $-D_z(p)$ is also decreasing with respect to $p_{z'}$ for $z' \neq z$.

The excess supply function Q is defined with

$$Q_z(p) = S_z(p) - D_z(p).$$

It is clearly continuous, and it is a Z-function as the sum of two Z-functions. Let's show that it is also strongly nonreversing, and therefore an M-function. The aggregates of $Q(p)$ are

$$\sum_z Q_z(p) = \sum_{x \in X} n_x \frac{\sum_z \exp(p_z - c_{xz})}{1 + \sum_z \exp(p_z - c_{xy})} - \sum_{y \in Y} m_y \frac{\sum_z \exp(a_{yz} - p_z)}{1 + \sum_z \exp(a_{yz} - p_z)},$$

which we see are strictly isotone in p: $p \geq p'$ implies $\sum_z Q_z(p) \geq \sum_z Q_z(p')$, and if $p \neq p'$ then the inequality is strict. Suppose then that we have $p \geq p'$ and $Q(p) \leq Q(p')$. Because aggregates are isotone, we must have $Q(p) = Q(p')$. Furthermore, if we had $p \neq p'$, then we would have $\sum_z Q_z(p) > \sum_z Q_z(p')$. But this contradicts $Q(p) = Q(p')$, hence actually $p = p'$.

Finally, note that when all $p_z = k$ and $k \to +\infty$, $Q_z(p) \geq 0$; while when all $p_z = k$ and $k \to -\infty$ we have $Q_z(p) \leq 0$. As a result both a super- and a subsolution exist, and therefore Theorem 2.16 guarantees that the problem has a unique solution which can be reached as the limit of a Jacobi sequence starting from any sub- or supersolution.

2.6 References and Notes

This section draws upon the book by Ortega and Rheinboldt [1], the papers by Berry, Gandhi and Haile [4], Galichon, Kominers and Weber [5], and Galichon, Samuelson and Vernet [2]. It also builds on the working papers by Galichon and Léger [3], and Chen, Choo, Galichon and Weber [6].

3 Models of Matching with Transfers

In this section we shall look specifically at models which formulate as

$$\begin{cases} \sum_{y\in Y} M_{xy}\left(a_x, b_y\right) + M_{x0}\left(a_x\right) = n_x \\ \sum_{x\in X} M_{xy}\left(a_x, b_y\right) + M_{0y}\left(b_y\right) = m_y \end{cases}$$

where $M_{xy}\left(a_x, b_y\right)$ is continuous and increasing in a_x and b_y, and will stand for the number of matches between types x and y; and $M_{x0}\left(a_x\right)$ and $M_{0y}\left(b_y\right)$ are also continuous and will stand for the number of unmatched agents of type x and y, respectively.

At the end of this section we will show how this framework extends to the full assignment case, i.e. when $M_{x0}\left(a_x\right) = M_{0y}\left(b_y\right) = 0$. Optimal transport falls into this category, as seen with Sinkhorn's algorithm in Sect. 2.

3.1 Microfoundation of the Matching Model

We will consider as illustration a matching model between workers (CEOs) and firms with taxes. There are n_x workers of type $x \in X$ and m_y firms of type $y \in Y$. Assume that firms pay their worker a gross wage w, from which the worker receives the net wage $N\left(w\right)$. In typical progressive taxation fashion, N is increasing, concave, and piecewise affine. We can thus represent it as

$$N\left(w\right) = \min_{k=0,\ldots,K} \left(1-\tau_k\right)\left(w - w_k\right) \tag{16}$$

where τ_k is the marginal tax rate in the kth tax bracket: $\tau_0 = 0 < \tau_1 < \cdots < \tau_K$. If there are no taxes, then we simply have $N\left(w\right) = w$.

If matched with a firm y, a worker x gets utility

$$\alpha_{xy} + N\left(w_{xy}\right) + \sigma\varepsilon_y$$

where α_{xy} is worker x's monetary valuation of job y's amenities; $N(w_{xy})$ is the net wage of worker x working for firm y; and ε_y is a random utility. If unmatched, a worker x gets utility $\sigma\varepsilon_0$. Hence worker x's problem is to choose a firm y with

$$\max_{y \in Y} \left\{\alpha_{xy} + N\left(w_{xy}\right) + \sigma\varepsilon_y,\ \sigma\varepsilon_0\right\}.$$

We assume that (ε_y) is i.i.d. Gumbel. As a result, the expected indirect utility of a worker of type x is

$$u_x = \sigma \log\left(1 + \sum_y \exp\left(\frac{\alpha_{xy} + N\left(w_{xy}\right)}{\sigma}\right)\right)$$

and the probability that worker x picks firm y is

$$\frac{\mu_{xy}}{n_x} = \frac{\exp\left(\frac{\alpha_{xy}+N\left(w_{xy}\right)}{\sigma}\right)}{1 + \sum_{y'} \exp\left(\frac{\alpha_{xy'}+N\left(w'_{xy}\right)}{\sigma}\right)} = \exp\left(\frac{\alpha_{xy} + N\left(w_{xy}\right) - u_x}{\sigma}\right). \tag{17}$$

On firm y's side, the problem is to choose a worker x with

$$\max_{x \in X} \left\{\gamma_{xy} - w_{xy} + \sigma\eta_x,\ \sigma\eta_0\right\}$$

where γ_{xy} is the value of worker x for firm y, and η_x is the random utility. We assume that (η_x) is also i.i.d. Gumbel, so that the expected indirect utility of a firm of type y is

$$v_y = \sigma \log\left(1 + \sum_x \exp\left(\frac{\gamma_{xy} - w_{xy}}{\sigma}\right)\right)$$

and the probability that firm y picks worker x is

$$\frac{\mu_{xy}}{m_y} = \exp\left(\frac{\gamma_{xy} - w_{xy} - v_y}{\sigma}\right). \tag{18}$$

Lastly, the probabilities that worker x or firm y remains unmatched are respectively:

$$\frac{\mu_{x0}}{n_x} = \exp\left(-\frac{u_x}{\sigma}\right), \qquad \frac{\mu_{0y}}{m_y} = \exp\left(-\frac{v_y}{\sigma}\right). \tag{19}$$

Equations (17), (18) and (19), together with the feasibility conditions

$$\sum_{y \in Y} \mu_{xy} + \mu_{x0} = n_x, \qquad \sum_{x \in X} \mu_{xy} + \mu_{0y} = m_y, \tag{20}$$

form a system of equations with unknowns $\mu_{xy}, \mu_{x0}, \mu_{0y}, u_x, v_y$ and w_{xy}. Our strategy to solve this system will consist of three steps. First, we will eliminate w_{xy} in order to express μ_{xy} as a function of u_x and v_y only, using the *distance-to-frontier function*. Second, we will solve for u_x and v_y with Jacobi's algorithm. Finally, we will recover $\mu_{xy}, \mu_{x0}, \mu_{0y}$ and w_{xy} from u_x and v_y.

3.2 Distance-to-Frontier Function

The wage w_{xy} is a way to transfer systematic utility from one partner to the other. After transfer, the systematic utilities of both partners in a match xy are

$$\begin{cases} \text{for } x,\ U_{xy} = \alpha_{xy} + N\left(w_{xy}\right) \\ \text{for } y,\ V_{xy} = \gamma_{xy} - w_{xy}. \end{cases}$$

More generally, we will consider models in which (U, V) belongs to a feasible set $\mathcal{F}_{xy}$, and we shall make two assumptions on $\mathcal{F}_{xy}$. The first one, *free disposal*, means that agents can dispose of utility at will.

Assumption 3.1 (Free disposal) If $(U, V) \in \mathcal{F}_{xy}$, then for any $U' \leq U$ and $V' \leq V$ we have $(U', V') \in \mathcal{F}_{xy}$.

The second one, *scarcity*, means that one partner cannot have an arbitrarily large level of utility without a negative effect on the utility of the other partner.

Assumption 3.2 (Scarcity) If (U^n) and (V^n) are two sequences such that $U^n \to +\infty$ and V^n is bounded below, then for N large enough $(U^n, V^n) \notin \mathcal{F}_{xy}$ for $n \geq N$; and similarly for U^n bounded below and $V^n \to +\infty$.

We will describe the feasible set $\mathcal{F}_{xy}$ using the *distance-to-frontier function*:

$$D_{xy}(U, V) = \min\left\{t \in \mathbb{R} : (U - t, V - t) \in \mathcal{F}_{xy}\right\}. \tag{21}$$

This function encodes the feasible set in the following sense. If the utility profile (U, V) is outside the feasible set, then the distance is positive: $D_{xy}(U, V) > 0$. Conversely, if (U, V) is in the interior of the feasible set, then the distance is negative: $D_{xy}(U, V) < 0$. Only when (U, V) is exactly on the frontier is the distance zero.

We mention two other useful properties of the distance-to-frontier function, before looking at this function on two examples.

Proposition 3.3 *(i) For F_1 and F_2 two elementary sets,*

$$D_{F_1 \cap F_2} = \max\left\{D_{F_1}, D_{F_2}\right\} \quad \textit{and} \quad D_{F_1 \cup F_2} = \min\left\{D_{F_1}, D_{F_2}\right\}.$$

(ii) Translation invariance: $D(U + t, V + t) = t + D(U, V)$.

Example 3.4 (Transferable utility) *In this case*

$$\mathcal{F}_{xy} = \left\{ (U, V) \in \mathbb{R}^2 : U + V \leq \Phi_{xy} \right\},$$

so the minimum t such that $(U - t, V - t) \in \mathcal{F}_{xy}$ verifies $(U - t) + (V - t) = \Phi_{xy}$, hence

$$D_{xy}(U, V) = \frac{U + V - \Phi_{xy}}{2}.$$

Example 3.5 (Piecewise linear taxes) *In this case $U \leq \alpha_{xy} + N\left(w_{xy}\right)$ and $V \leq \gamma_{xy} - w_{xy}$, so that*

$$\begin{aligned}
\mathcal{F}_{xy} &= \left\{ (U, V) \in \mathbb{R}^2 : N\left(\gamma_{xy} - V\right) \geq U - \alpha_{xy} \right\} \\
&= \left\{ (U, V) : \min_{k=0,\ldots,K} (1 - \tau_k)(\gamma_{xy} - V - w_k) \geq U - \alpha_{xy} \right\} \\
&= \bigcap_k \left\{ (U, V) : (1 - \tau_k)(\gamma_{xy} - V - w_k) \geq U - \alpha_{xy} \right\}.
\end{aligned}$$

Hence $D_{xy}(U, V) = \max_k D_{xy}^k(U, V)$, where

$$D_{xy}^k(U, V) = \frac{U - \alpha_{xy} + (1 - \tau_k)(V - \gamma_{xy} + w_k)}{2 - \tau_k}.$$

3.3 *Matching Equilibrium*

Now back to the matching model, recall that we had

$$\frac{\mu_{xy}}{n_x} = \exp\left(\frac{U_{xy} - u_x}{\sigma}\right), \qquad \frac{\mu_{xy}}{m_y} = \exp\left(\frac{V_{xy} - v_y}{\sigma}\right),$$

where U_{xy} and V_{xy} denote the systematic parts of utility in a match xy. Solving for U_{xy} and V_{xy} yields

$$U_{xy} = u_x + \sigma \log \frac{\mu_{xy}}{n_x}, \qquad V_{xy} = v_y + \sigma \log \frac{\mu_{xy}}{m_y}.$$

We assume that the market is large, so that every type of match xy must be occurring. Then $D_{xy}\left(U_{xy}, V_{xy}\right) = 0$ for all xy. Replacing U_{xy} and V_{xy} by their expressions above, we get

$$D_{xy}\left(u_x + \sigma \log \frac{\mu_{xy}}{n_x}, v_y + \sigma \log \frac{\mu_{xy}}{m_y}\right) = 0.$$

The translation invariance property yields

$$\sigma \log \mu_{xy} + D_{xy}\left(u_x - \sigma \log n_x, v_y - \sigma \log m_y\right) = 0.$$

Thus we have eliminated w_{xy} by combining equations (17) and (18), and we are able to express μ_{xy}, μ_{x0} and μ_{0y} as functions of u_x and v_y only:

$$\mu_{xy} = \exp\left(-\frac{D_{xy}\left(u_x - \sigma \log n_x, v_y - \sigma \log m_y\right)}{\sigma}\right),$$

$$\mu_{x0} = \exp\left(-\frac{u_x - \sigma \log n_x}{\sigma}\right), \qquad \mu_{0y} = \exp\left(-\frac{v_y - \sigma \log m_y}{\sigma}\right).$$

Next we want to solve for u_x and v_y using Jacobi's algorithm. Introduce $p_x = -(u_x - \sigma \log n_x)$ and $p_y = v_y - \sigma \log m_y$, and substitute the expressions of μ_{xy}, μ_{x0} and μ_{0y} above in the feasibility conditions (20) to obtain the system:

$$\begin{cases} \sum_y \exp\left(-D_{xy}\left(-p_x, p_y\right)/\sigma\right) + \exp\left(p_x/\sigma\right) = n_x \\ \sum_x \exp\left(-D_{xy}\left(-p_x, p_y\right)/\sigma\right) + \exp\left(-p_y/\sigma\right) = m_y. \end{cases} \tag{22}$$

We define the function Q using

$$\begin{cases} Q_x\left(p\right) = \sum_y \exp\left(-D_{xy}\left(-p_x, p_y\right)/\sigma\right) + \exp\left(p_x/\sigma\right) - n_x \\ Q_y\left(p\right) = -\sum_x \exp\left(-D_{xy}\left(-p_x, p_y\right)/\sigma\right) - \exp\left(-p_y/\sigma\right) + m_y, \end{cases}$$

so that the system (22) rewrites as $Q\left(p\right) = 0$. One can then check that Q is a continuous M-function (it is a Z-function with strictly isotone aggregates).

We also have:

Proposition 3.6 *The problem $Q\left(p\right) = 0$ has a sub- and a supersolution.*

Proof. To get a supersolution, first set p_x such that $\exp\left(p_x/\sigma\right) \geq n_x$, so that $Q_x\left(p_x, p_{-x}\right) \geq 0$ for all p_{-x}. Second, we show that for all xy we must have $D_{xy}(-p_x, p_y) \to +\infty$ when $p_y \to +\infty$. Suppose this is not the case: then there exists an increasing sequence (p_y^n) such that $p_y^n \to +\infty$ but $D_{xy}(-p_x, p_y^n) \leq K$ for some constant K. By translation invariance, $D_{xy}(-p_x - K, p_y^n - K) \leq 0$, i.e. $(-p_x - K, p_y^n - K) \in \mathcal{F}_{xy}$ for all n. But since $-p_x - K$ is constant (so bounded below) and $p_y^n - K \to +\infty$, this contradicts the scarcity assumption. Hence for p_y large enough, $\sum_x \exp\left(-D_{xy}\left(-p_x, p_y\right)/\sigma\right) + \exp\left(-p_y/\sigma\right) \leq m_y$, i.e. $Q_y\left(p\right) \geq 0$. A subsolution can be found in a similar manner. □

Hence by Theorem 2.16, the problem has a unique solution. Finally, since the problem has a solution for any n_x and m_y the function Q must be surjective, and therefore any Jacobi sequence converges towards the solution according to Theorem 2.17.

We conclude this discussion with a comment on the optimization structure. If Q were a gradient, one could write $Q_z = \partial F/\partial p_z$ for some function F and thus

$$\frac{\partial Q_y}{\partial p_x} = \frac{\partial^2 F}{\partial p_x \partial p_y} = \frac{\partial Q_x}{\partial p_y}.$$

But we have

$$\frac{\partial Q_x(p)}{\partial p_y} = -\exp\left(-\frac{D_{zy}(-p_x, p_y)}{\sigma}\right)\frac{\partial_V D_{xy}(-p_x, p_y)}{\sigma}$$

$$\frac{\partial Q_y(p)}{\partial p_x} = -\exp\left(-\frac{D_{zy}(-p_x, p_y)}{\sigma}\right)\frac{\partial_U D_{xy}(-p_x, p_y)}{\sigma}$$

which is not symmetric unless

$$\partial_U D_{xy}(U, V) = \partial_V D_{xy}(U, V).$$

This only happens in the optimal transport case, when $D_{xy}(U, V) = (U + V - \Phi_{xy})/2$.

3.4 Full Assignment Case

Here we assume that $\sum_x n_x = \sum_y m_y$. In the previous setting, define the map $Q : \mathbb{R}^{X \cup Y \setminus \{y_0\}} \to \mathbb{R}^{X \cup Y \setminus \{y_0\}}$ by

$$\begin{cases} Q_x(p) = \sum_y \exp\left(-D_{xy}(-p_x, p_y)\right) - n_x, & x \in X \\ Q_y(p) = -\sum_x \exp\left(-D_{xy}(-p_x, p_y)\right) + m_y, & y \in Y \setminus \{y_0\} \end{cases}$$

where we have normalized $p_{y_0} = \pi$. To ensure that Q is responsive in this setting (and therefore that Jacobi updates are well-defined) we will make an additional assumption on the feasible sets $\mathcal{F}_{xy}$:

Assumption 3.7 (Strong transferability) For all U there exists V such that $(U, V) \in \mathcal{F}_{xy}$; and conversely, for all V there exists U such that $(U, V) \in \mathcal{F}_{xy}$.

Under this assumption, any arbitrarily high level of utility is attainable for a partner within a match, as long as the other partner is willing to give up enough utility.

Chen, Choo, Galichon and Weber [6] show:

Theorem 3.8 *The equation $Q(p) = 0$ has a solution.*

Proof. The proof is constructive and is done in three steps. First, we look for a supersolution to initialize Jacobi's algorithm. Second, we show that Jacobi updates are well-defined. Finally, we show that the Jacobi sequence cannot diverge.

Step 1. Look for a supersolution p^0. We have

$$Q_x(p) = \sum_y \exp\left(-D_{xy}(-p_x, p_y)\right) - n_x \geq \exp\left(-D_{xy_0}(-p_x, \pi)\right) - n_x.$$

By strong transferability, there exists U_x such that $(U_x, \pi + \log n_x) \in \mathcal{F}_{xy}$, i.e. $D_{xy}(U_x, \pi + \log n_x) \leq 0$. Taking $p_x^0 = \log n_x - U_x$, we have $0 \geq D_{xy}(-p_x^0 + \log n_x, \pi + \log n_x) = D_{xy}(-p_x^0, \pi) + \log n_x$, i.e. $\exp\left(-D_{xy}(-p_x^0, \pi)\right) \geq n_x$, and therefore $Q_x\left(p_x^0, p_{-x}\right) \geq 0$ for all p_{-x}. Next, by scarcity we have $D_{xy}(-p_x^0, p_y) \to +\infty$ as $p_y \to +\infty$ (see the argument in the proof of Proposition 3.6) so we can set p_y^0 large enough such that

$$\sum_x \exp\left(-D_{xy}(-p_x^0, p_y^0)\right) \leq m_y,$$

hence $Q_y\left(p^0\right) \geq 0$ as well, so we have a supersolution.

Step 2. We now show that the Jacobi update from a supersolution is well-defined. The function $p_x \mapsto \sum_y \exp\left(-D_{xy}(-p_x, p_y^0)\right) - n_x$ is continuous, nonnegative for $p_x = p_x^0$, and by scarcity it converges to $-n_x < 0$ when $p_x \to -\infty$, hence p_x^1 is well-defined.

Now consider the function $p_y \mapsto -\sum_x \exp\left(-D_{xy}(-p_x^0, p_y)\right) + m_y$. It is continuous, and nonnegative for $p_y = p_y^0$. Choose an x arbitrarily, then by strong transferability, there exists V_y such that $(-p_x^0 + \log m_y, V_y) \in \mathcal{F}_{xy}$. We have $0 \geq D_{xy}(-p_x^0 + \log m_y, V_y) = D_{xy}(-p_x^0, V_y - \log m_y) + \log m_y$, thus $\exp\left(-D_{xy}(-p_x^0, V_y - \log m_y)\right) \geq m_y$, and therefore the function above is negative for $p_y = V_y - \log m_y$, hence p_y^1 is also well-defined.

By induction, p^{t+1} is well-defined for all t since p^t is also a supersolution.

Step 3. Let us show that the Jacobi sequence initialized at the supersolution p^0 cannot diverge. Q is a Z-function with isotone aggregates, therefore it is an M_0-function, and thus the Jacobi sequence (p^t) starting from p^0 is a decreasing sequence of supersolutions. Hence, either it converges to a solution, or it is unbounded.

However, because p^t is a supersolution, we have

$$\sum_x \exp\left(-D_{xy_0}(-p_x^t, \pi)\right) - m_{y_0} = \sum_{z \in X \cup Y \setminus \{y_0\}} Q_z\left(p^t\right) \geq 0$$

thus all the p_x^t cannot go to $-\infty$. Letting x^* be such that $p_{x^*}^t \to p_{x^*}^\infty > -\infty$, we have

$$\exp\left(-D_{x^*y}(-p_{x^*}^t, p_y^t)\right) \leq \sum_x \exp\left(-D_{xy}(-p_x^t, p_y^t)\right) \leq m_y$$

thus all p_y^t remain bounded below as well. Finally, we have

$$\sum_y \exp\left(-D_{xy}(-p_x^t, p_y^t)\right) \geq n_x$$

and thus all the p_x^t remain bounded below too. Hence the sequence converges. □

Uniqueness of the solution is not guaranteed under the strong transferability assumption, because then Q is a M_0-function. To obtain uniqueness, one may ask an additional requirement on the feasible sets, which ensures that Q is a M-function and therefore that Theorem 2.16 applies. For instance:

Assumption 3.9 (Strong local transferability) For any $(U, V) \in \mathcal{F}_{xy}$ and any $\varepsilon > 0$, the open ball of center (U, V) and of radius ε contains both a point (U', V') such that $U' > U$, and a point $(U'', V'') \in \mathcal{F}_{xy}$ such that $V'' > V$.

This assumption rules out horizontal or vertical slopes on the frontier of $\mathcal{F}_{xy}$, which ensures that Q is an M-function.

References and Notes

This section draws upon Galichon, Kominers and Weber [5]. The case with full assignment is covered in Chen, Choo, Galichon and Weber [6].

4 Models of Matching Without Transfers

In the previous section, utility could be (imperfectly) transferred within matches by adjusting the wage that the firm was paying to the worker. In this section, we consider instead models in which each match leads to a fixed amount of utility for each partner.

4.1 Introduction

We consider a labor market with fixed wages. Assume that $\mathcal{I}$ is the set of (individual) workers, and $\mathcal{J}$ is the set of (individual) firms. A worker i matched with firm j gets utility α_{ij}, and similarly, a firm j matched with worker i gets utility γ_{ij}. Unassigned agents (workers or firms) get utility 0.

We make an important assumption:

Assumption 4.1 (No indifference) $\alpha_{ij} \neq 0$ and $\alpha_{ij} \neq \alpha_{ij'}$ for $j \neq j'$, and similarly, $\gamma_{ij} \neq 0$ and $\gamma_{ij} \neq \gamma_{i'j}$ for $i \neq i'$.

We are looking at a model of one-to-one matching. A matching $\mu = (\mu_{ij})$ is defined such that $\mu_{ij} \in \{0, 1\}$ is equal to 1 if i and j are matched, and 0 otherwise. μ_{i0} or μ_{0j} are respectively equal to 1 if i or j are unassigned. The condition for μ to be feasible is therefore

$$\begin{cases} \sum_j \mu_{ij} + \mu_{i0} = 1 \\ \sum_i \mu_{ij} + \mu_{0j} = 1. \end{cases} \tag{23}$$

Example 4.2 *Consider two workers and one firm. Agents have a utility 1 if they are matched, and a utility 0 if they are not matched. We see that both workers want to be matched with the firm; however, the firm can only match with one worker. Unlike the models with transfers from Sect. 3, there is no wage to clear the market. As a result, at least one worker will not get their first choice.*

As seen in the previous example, we can follow two alternative approaches. The first one requires to give up on the idea of equilibrium, which implies that individuals get their first choice. In this case, we need a central planner to solve the market using an algorithm: this is Gale and Shapley's notion of a *stable matching*, which can be computed by the *deferred acceptance algorithm* (see [7] and [8]).

The second method instead introduces a numéraire that cannot be transferred, like waiting times, to clear the market: this is the approach introduced in [2]. This approach provides *equilibrium matchings* that can be computed by the *deferred acceptance with Lagrange multipliers (DALM) algorithm*.

We will review both approaches in that order, and then study their relationships. In doing so we will also cover Adachi's reformulation of Gale and Shepley's stable matching problem, which allows to use the Gauss–Seidel algorithm (closely related to Jacobi's algorithm) to solve the problem.

4.2 Stable Matchings

Introduce u_i and v_j the respective payoffs obtained by worker $i \in \mathcal{I}$ and firm $j \in \mathcal{J}$ in the outcome of the game.

Note that if there were a pair i and j for which $u_i < \alpha_{ij}$ and $v_j < \gamma_{ij}$, then by matching together i and j could both obtain more utility than what is guaranteed to them in the outcome of the game. This outcome would then not be stable, and i and j would form a *blocking pair*. Similarly, if $u_i < 0$, then i would be better off remaining unmatched and getting utility 0 rather than taking the outcome payoff u_i. A similar result holds on the other side of the market, and as a result we shall require that for all i and j,

$$\max\left(u_i - \alpha_{ij}, v_j - \gamma_{ij}\right) \geq 0,$$
$$u_i \geq 0, \ v_j \geq 0,$$

which are called the *stability conditions*.

Finally, if i and j actually match, we expect u_i to be equal to α_{ij} and v_j to be equal to γ_{ij}. Similarly, if i or j respectively remains unmatched, then u_i or v_j should be 0. We call this a *strong complementarity condition*, for reasons that will become clear later on.

To summarize, an outcome (μ, u, v) is *stable* if:

(i) μ is a feasible matching: $\mu_{ij} \in \{0, 1\}$ and

$$\textstyle\sum_j \mu_{ij} + \mu_{i0} = 1 \tag{24}$$

$$\textstyle\sum_i \mu_{ij} + \mu_{0j} = 1, \tag{25}$$

(ii) the stability conditions hold, that is

$$\max\left(u_i - \alpha_{ij}, v_j - \gamma_{ij}\right) \geq 0, \tag{26}$$

$$u_i \geq 0,\ v_j \geq 0, \tag{27}$$

(iii) strong complementarity holds, that is

$$\mu_{ij} > 0 \implies u_i = \alpha_{ij},\ v_j = \gamma_{ij} \tag{28}$$
$$\mu_{i0} > 0 \implies u_i = 0 \tag{29}$$
$$\mu_{0j} > 0 \implies v_j = 0. \tag{30}$$

Because of the strong complementarity conditions, one can deduce the utilities u and v of a stable outcome from its matching μ. This is in contrast with models of matching with transfers, in which different utility profiles can be compatible with the same matching. For this reason, in models of matching without transfers we sometimes refer to *stable matchings* instead of stable outcomes.

The standard method to find such a stable matching is Gale and Shapley's deferred acceptance algorithm. The idea is as follows. We keep track of a set of "available" offers that can be made by workers to firms; this is initially unconstrained, which means that any worker can initially make an offer to any firm. At each round, each worker makes an offer to their preferred firm among the set of those that are available to them. If a firm receives several offers, it keeps its favorite and rejects the others. Offers that have been rejected are no longer available. The algorithm then iterates until no offer is rejected.

Formally, let's define $\mathcal{A}^t(i) \subseteq \mathcal{J}$ as the set of firms available to worker i at time t, $\mathcal{P}^t(i) \subseteq \mathcal{J}$ as the set of firms to whom worker i makes an offer at time t (either a singleton or empty), and $\mathcal{K}^t(j) \subseteq I$ as the set of offers kept by firm j at the end of round t (also either a singleton or empty). Assume that at $t = 0$ all firms are available to anyone: $\mathcal{A}^0(i) = \mathcal{J}$.

Deferred acceptance algorithm (Gale and Shapley [7])

Iterate over $t \geq 0$:

1. Proposal phase:

$$\mathcal{P}^t(i) = \arg\max_j \left\{\alpha_{ij} : j \in \mathcal{A}^t(i)\right\}$$

or $\mathcal{P}^t(i) = \emptyset$ if the max is negative or $\mathcal{A}^t(i)$ is empty,

2. Disposal phase:

$$\mathcal{K}^t(j) = \arg\max_i \left\{\gamma_{ij} : j \in \mathcal{P}^t(i)\right\}$$

or $\mathcal{K}^t(j) = \emptyset$ if the max is negative or $j \notin \bigcup_i \mathcal{P}^t(i)$,

3. Adjustment phase:

$$\mathcal{A}^{t+1}(i) = \mathcal{A}^t(i) \setminus \left(\mathcal{P}^t(i) \setminus (\mathcal{K}^t)^{-1}(i)\right)$$

where $(\mathcal{K}^t)^{-1}(i)$ is the set of firms who kept worker i's offer,

until $\mathcal{A}^{t+1}(i) = \mathcal{A}^t(i)$.

Theorem 4.3 *The deferred acceptance algorithm converges to a stable matching.*

Proof. The proof is deferred. □

4.3 Adachi's Formulation

Adachi's formulation stems from a simple idea. As argued above, in a stable matching individuals do not always get their first choice, since there is no numéraire to clear the market. However, for each worker, one can define the *consideration set* as the set of firms that are willing to match with that worker; in a stable matching, each worker is matched with their preferred firm among the consideration set. Similarly on the other side of the market, each firm is matched with their preferred worker among its consideration set. Adachi's theorem in [9] states that this is actually a necessary and sufficient condition for a stable matching.

Theorem 4.4 (Adachi [9]) *The outcome (μ, u, v) is stable if and only if*

$$\begin{cases} u_i = \max_j \left\{\alpha_{ij}, 0 : v_j \leq \gamma_{ij}\right\} \\ v_j = \max_i \left\{\gamma_{ij}, 0 : u_i \leq \alpha_{ij}\right\}. \end{cases} \tag{31}$$

Note that we use the shortened notation $\max_j \left\{\alpha_{ij}, 0 : v_j \leq \gamma_{ij}\right\}$ to mean $\max(0, \max_j \left\{\alpha_{ij} : v_j \leq \gamma_{ij}\right\})$.

We can make some remarks. First, this theorem yields a formulation of stable matchings as the fixed point of some operator – we will come back to this below. Second, the mapping $v \mapsto u$ is antitone with respect to v, and the mapping $u \mapsto v$ is antitone with respect to u. Lastly, the function $u_i(v) = \max_j \left\{\alpha_{ij} : v_j \leq \gamma_{ij}\right\}$ is reminiscent of the convex conjugate function $v_i^*(v) = \max_j \left\{\Phi_{ij} - v_j\right\}$. However, $u = v^*$ and $v = u^*$ is not sufficient for (u, v) to be solution to the dual optimal transport problem.

Proof. For simplicity, we present the proof when $|\mathcal{I}| = |\mathcal{J}|$ and all α and γ are positive, so that in a stable outcome everyone must be matched. For the direct implication, assume that (μ, u, v) is stable. Looking for a contradiction, assume that (31) does not hold: without loss of generality,

$$u_i \neq \max_j \left\{\alpha_{ij} : v_j \leq \gamma_{ij}\right\} \quad \text{for some } i.$$

Worker i is matched with $J(i)$, so that $u_i = \alpha_{iJ(i)}$ and $v_{J(i)} = \gamma_{iJ(i)}$. As a result

$$\max_j \left\{\alpha_{ij} : v_j \leq \gamma_{ij}\right\} \geq \alpha_{iJ(i)} = u_i$$

thus we have $\max_j \left\{\alpha_{ij} : v_j \leq \gamma_{ij}\right\} > u_i = \alpha_{iJ(i)}$. Denote $j^* \neq J(i)$ the corresponding maximizer, we have $v_{j^*} \leq \gamma_{ij^*}$ and $\alpha_{ij^*} > u_i$. But no indifference implies $v_{j^*} < \gamma_{ij^*}$, therefore i and j^* form a blocking pair. Contradiction, thus (31) in fact holds.

Conversely, assume that the equalities (31) hold. Define $\mu_{ij} = 1$ iff $u_i = \alpha_{ij}$. Since $u_i = \alpha_{ij}$, j must be in the feasible set for i, i.e. $v_j \leq \gamma_{ij}$. Furthermore, i is such that $u_i \leq \alpha_{ij}$, thus $v_j \geq \gamma_{ij}$. As a result, $v_j = \gamma_{ij}$. By no indifference, it also means that the matching μ thus defined is feasible.

Let us show that (μ, u, v) is Gale–Shapley stable. Assume ij is a blocking pair, so that $\alpha_{ij} > u_i$ and $\gamma_{ij} > v_j$. But $\gamma_{ij} > v_j$ implies $v_j \leq \gamma_{ij}$, thus $u_i \geq \alpha_{ij}$, which is a contradiction. □

Galichon and Léger [3] build on Adachi's formulation to rewrite the stability problem as a model with substitutes, specifically in the form $Q(p) = 0$ with Q an M_0-function.

Use the change-of-sign trick in (31): set $p_i = -u_i$ and $p_j = v_j$, and define Adachi's map T as

$$\begin{cases} T_i(p) = \min_j \left\{-\alpha_{ij}, 0 : \gamma_{ij} \geq p_j\right\} \\ T_j(p) = \max_i \left\{\gamma_{ij}, 0 : p_i \geq -\alpha_{ij}\right\}. \end{cases}$$

Clearly T is an isotone map, and by Adachi's theorem we know that (μ, u, v) is stable if and only if $p = (-u, v)$ satisfies the fixed-point equation

$$p = T(p).$$

Even though $p - T(p)$ is a Z-function, it is not an M_0-function in general. We set instead

$$\begin{cases} M_{ij}(p) = 1\left\{p_i \geq -\alpha_{ij}, \gamma_{ij} \geq p_j\right\} \\ M_{i0}(p) = 1\left\{p_i \geq 0\right\} \\ M_{0j}(p) = 1\left\{0 \geq p_j\right\} \end{cases}$$

and we define the function Q with

$$\begin{cases} Q_i(p) = \sum_{j \in \mathcal{J}} M_{ij}(p) + M_{i0}(p) - 1 \\ Q_j(p) = -\sum_{i \in \mathcal{I}} M_{ij}(p) - M_{0j}(p) + 1. \end{cases}$$

Although Q is not continuous, we have:

Theorem 4.5 (Galichon and Léger [3])

(i) Q is an M_0-function.
(ii) Adachi's map T is the coordinate update operator associated with Q.

As a result, $T(p) = p$ if and only if $Q(p) = 0$, and thus the set of zeroes of Q is a sublattice of $\mathbb{R}^n$ (c.f. corollary of Theorem 2.15). Hence:

***Corollary* (Knuth [10])** *Given two stable outcomes (μ, u, v) and (μ', u', v'), define*

$$\begin{cases} (\mu \wedge_{\mathcal{I}} \mu')_{ij} = 1\left\{u_i \leq u'_i\right\} \mu_{ij} + 1\left\{u_i > u'_i\right\} \mu'_{ij} \\ (\mu \vee_{\mathcal{I}} \mu')_{ij} = 1\left\{u_i > u'_i\right\} \mu_{ij} + 1\left\{u_i \leq u'_i\right\} \mu'_{ij}. \end{cases} \tag{32}$$

Then $\mu \wedge_{\mathcal{I}} \mu'$ and $\mu \vee_{\mathcal{I}} \mu'$ are stable matchings, and

$$\begin{cases} \mu \wedge_{\mathcal{I}} \mu' = \mu \vee_{\mathcal{J}} \mu' \\ \mu \vee_{\mathcal{I}} \mu' = \mu \wedge_{\mathcal{J}} \mu'. \end{cases} \tag{33}$$

As a result, the lattice of stable matchings has an element which is preferred by all the i's and least liked by the j's; and an element which is preferred by all the j's and least liked by the i's.

In other words, workers i and firms j have *opposite interests*: what is better for workers is worse for firms and conversely.

Adachi's algorithm is nothing else than Gauss–Seidel's algorithm applied to finding the zeroes of Q. (Gauss–Seidel's algorithm is closely related to Jacobi's algorithm, the difference being that coordinate updates are made sequentially with Gauss–Seidel, instead of in parallel with Jacobi.)

Adachi's algorithm

Set p^0 low enough, for instance $p_i^0 = \min_j \left\{-\alpha_{ij}, 0\right\}$ and $p_j^0 = \min_i \left\{\gamma_{ij}, 0\right\}$. Iterate over $t \geq 0$:

$$\begin{cases} p_i^{t+1} = \min_j \left\{-\alpha_{ij}, 0 : \gamma_{ij} \geq p_j^t\right\} \\ p_j^{t+1} = \max_i \left\{\gamma_{ij}, 0 : p_i^{t+1} \geq -\alpha_{ij}\right\} \end{cases} \tag{34}$$

until $p^{t+1} = p^t$.

Note that if it starts from a subsolution, Adachi's algorithm will return the $\mathcal{I}$-preferred stable matching, and if it starts from a supersolution, it will return the $\mathcal{J}$-preferred stable matching.

In Gale and Shapley's deferred acceptance algorithm, at each period u_i^t can only decrease by one step in the scale of the rankings induced by α_{ij}. Indeed, if the offer made by worker i is kept, then i keeps making the same offer; while if the offer made by i is turned down, then i will move to their next preferred firm. This may however be inefficient, as i is allowed to make an offer to a firm j who already has a dominating offer, and who will thus turn down i.

A more efficient version of deferred acceptance consists in modifying Gale and Shapley's algorithm in order to induce the workers i whose offer has been turned down to move to the next firm j in their preference list which does not already have an offer better than i. This is exactly what Adachi's algorithm does.

Formally, let us denote

$$\begin{cases} N_i(p) = \max_j \left\{-\alpha_{ij} : -\alpha_{ij} < p_i\right\} \\ N_j(p) = \max_i \left\{\gamma_{ij} : \gamma_{ij} < p_j\right\} \end{cases} \tag{35}$$

which looks for the match just below the match implied by p for the i's, and the match just above the matched implied by p for the j's.

Gale and Shapley's algorithm can be interpreted as

$$\begin{cases} p_i^{2t+1} = \min\left\{T_i\left(p^{2t}\right), N_i\left(p^{2t}\right)\right\} \\ p_j^{2t+2} = T_i\left(p^{2t+1}\right) \end{cases} \tag{36}$$

which is a "damped Gauss–Seidel" algorithm (where T is Adachi's map).

4.4 Equilibrium Matchings

The notion of equilibrium matchings was introduced by Galichon, Hsieh and Sylvestre [11]. Just like stable matchings, equilibrium matchings require to rule out blocking pairs, so the stability conditions will be the same. The only difference is that we shall not require that if i and j are matched, then $u_i = \alpha_{ij}$ and $v_j = \gamma_{ij}$, but instead that $u_i \leq \alpha_{ij}$ and $v_j \leq \gamma_{ij}$ with at most one of these inequalities being strict. That is, we are fine with burning utility on one side of the market provided that we don't burn it on both sides. Hence the strong complementarity condition (28) becomes a weak one, only requiring that

$$\mu_{ij} > 0 \implies \max\left(u_i - \alpha_{ij}, v_j - \gamma_{ij}\right) = 0. \tag{37}$$

An outcome (μ, u, v) is thus an *equilibrium matching* if it satisfies conditions (24)–(30), but replacing the strong complementarity condition (28) with (37).

Of course, if an outcome is stable, then it is an equilibrium matching. We have seen that stable outcomes exist, so why is the notion of an equilibrium matching useful? The reason is that we can define a notion of aggregate equilibrium matching, but not of aggregate stable matching.

We will now discard the no-indifference assumption by considering a population similar to the one we studied in the matching with transfers setting. Workers and firms are sorted into types $x \in \mathcal{X}$ and $y \in \mathcal{Y}$ respectively, so that there are n_x identical workers of type x, and m_y identical firms of type y. The value of the match xy is α_{xy} for a worker x, and γ_{xy} for a firm y. Hence workers x are indifferent between all firms y, and reciprocally. Example 4.2 showed that in this case, aggregate stable matchings may not necessarily exist since two workers x might end up with different utilities depending on which firm they are matched with, when we want to be able to define u_x the ex-post utility for all workers x. In contrast, aggregate equilibrium matchings do exist.

Definition 4.6 *An outcome* (μ, u, v) *is an* aggregate equilibrium matching *if:*

(i) μ *is a feasible matching:* $\mu_{xy} \geq 0$*, and*

$$\textstyle\sum_y \mu_{xy} + \mu_{x0} = n_x \tag{38}$$

$$\textstyle\sum_x \mu_{xy} + \mu_{0y} = m_y, \tag{39}$$

(ii) the stability conditions hold, that is

$$\max\left(u_x - \alpha_{xy}, v_y - \gamma_{xy}\right) \geq 0 \tag{40}$$

$$u_x \geq 0,\ v_y \geq 0, \tag{41}$$

(iii) weak complementarity holds, that is

$$\mu_{xy} > 0 \implies \max\left(u_x - \alpha_{xy}, v_y - \gamma_{xy}\right) = 0 \tag{42}$$

$$\mu_{x0} > 0 \implies u_x = 0 \tag{43}$$

$$\mu_{0y} > 0 \implies v_y = 0. \tag{44}$$

It is clear that this definition coincides with the notion of (individual) equilibrium matching. We have just replaced the margins and allowed μ_{xy} to take other values than just 0 or 1. This is part of the theory seen in Sect. 3 with the distance function

$$D_{xy}(U, V) = \max\left(U - \alpha_{xy}, V - \gamma_{xy}\right). \tag{45}$$

To compute aggregate equilibrium matchings, [11] considers an aggregate version of Gale and Shapley's deferred acceptance algorithm. Let $\mu_{xy}^{A,t}$ denote the number of positions of type y available to workers of type x at time t. At $t = 0$ all positions are open: $\mu_{xy}^{A,0} = \min\left\{n_x, m_y\right\}$.

Deferred Acceptance with Lagrange multipliers (DALM)

Define $\mu_{xy}^{A,0} = \min\left\{n_x, m_y\right\}$, and iterate over $t \geq 0$:

1. Proposal phase:

$$\mu^{P,t} \in \arg\max_{\mu} \sum_{xy} \mu_{xy}\alpha_{xy}$$
$$s.t. \sum_{y} \mu_{xy} \leq n_x$$
$$\mu_{xy} \leq \mu_{xy}^{A,t}$$

2. Disposal phase:

$$\mu^{K,t} \in \arg\max_{\mu} \sum_{xy} \mu_{xy}\gamma_{xy}$$
$$s.t. \sum_{x} \mu_{xy} \leq m_y$$
$$\mu_{xy} \leq \mu_{xy}^{P,t}$$

3. Adjustment phase: $\mu_{xy}^{A,t+1} = \mu_{xy}^{A,t} - (\mu_{xy}^{P,t} - \mu_{xy}^{K,t})$

until $\mu_{xy}^{A,t+1} = \mu_{xy}^{A,t}$.

In the proposal and disposal phase, there are Lagrange multipliers $\tau_{xy}^{\alpha} \geq 0$ and $\tau_{xy}^{\gamma} \geq 0$ which are such that the perceived utilities are respectively $\alpha_{xy} - \tau_{xy}^{\alpha}$ for workers, and $\gamma_{xy} - \tau_{xy}^{\gamma}$ for firms. We can then show that τ_{xy}^{α} is increasing and τ_{xy}^{γ} is decreasing.

We have the following result:

Theorem 4.7 (Galichon, Hsieh and Sylvestre [11]) *Aggregate equilibrium matchings exist, and DALM returns one of them.*

The proof relies on novel comparative statics for *exchangeable functions* – which are not covered by Topkis' and Veinott's monotone comparative statics theory.

4.5 Application: Housing Market with Rent Control

Consider a housing market in which the rent is capped. Renters i are categorized according to family size $x \in X$. Houses j are categorized according to the number of rooms $y \in Y$. We assume that the rent for a house y is capped at r_y and that houses are under-supplied in the market, so that in practice r_y is the going rent for any house y.

The utility of a family i of size x renting a house y is $a_{xy} - r_y + \varepsilon_{iy}$, where a_{xy} represents the monetary value of the amenities of a house y for a family of size x, and ε_{iy} is a random utility component. If unmatched, the family gets utility ε_{i0}. The utility of a landlord j who rents out a house y to a family of size x is $r_y - c_{xy} + \eta_{xj}$, where c_{xy} is the maintenance cost which depends on the family size and house type, and η_{xj} is also random. If unmatched, the landlord gets utility η_{0j}.

We can let $\alpha_{xy} = a_{xy} - r_y$ and $\gamma_{xy} = r_y - c_{xy}$, and the analysis follows the same steps as in Sect. 3. The respective problems of family i and landlord j are

$$\max_{y} \{\alpha_{xy} - t_{xy} + \varepsilon_{iy}, \varepsilon_{i0}\}, \qquad \max_{x} \{\gamma_{xy} - s_{xy} + \eta_{xj}, \eta_{0j}\}$$

where $t_{xy}, s_{xy} \geq 0$ are the "burnt utilities" on each side of the match.

Assuming the random utility components are i.i.d. Gumbel, the expected indirect utilities for a family x and a landlord y are

$$u_x = \log\left(1 + \sum_y \exp\left(\alpha_{xy} - t_{xy}\right)\right), \qquad v_y = \log\left(1 + \sum_x \exp\left(\gamma_{xy} - s_{xy}\right)\right)$$

and the probabilities that a family of type x picks a house y, and that a landlord y chooses a family x, are respectively

$$\frac{\mu_{xy}}{n_x} = \exp\left(\alpha_{xy} - t_{xy} - u_x\right), \qquad \frac{\mu_{xy}}{m_y} = \exp\left(\gamma_{xy} - s_{xy} - v_y\right).$$

With the large market assumption, so that every type of match occurs in equilibrium, utility can only be burnt on one side of each match xy, so that

$$\min(t_{xy}, s_{xy}) = 0.$$

Substituting t_{xy} and s_{xy} from the equations above, we obtain

$$\log \mu_{xy} = \min\left(\alpha_{xy} - u_x + \log n_x +, \gamma_{xy} - v_y + \log m_y\right).$$

This corresponds exactly to what we had in Sect. 3, but adapted to the distance-to-frontier function in (45). Using $\mu_{x0} = \exp(-u_x + \log n_x)$ and $\mu_{0y} = \exp(-v_y + \log m_y)$, this rewrites as

$$\mu_{xy} = \min\left\{\mu_{x0} \exp\left(\alpha_{xy}\right), \mu_{0y} \exp\left(\gamma_{xy}\right)\right\}.$$

Hence μ_{x0} and μ_{0y} must satisfy the system

$$\begin{cases} \mu_{x0} + \sum_y \min\left\{\mu_{x0} \exp\left(\alpha_{xy}\right), \mu_{0y} \exp\left(\gamma_{xy}\right)\right\} = n_x \\ \mu_{0y} + \sum_x \min\left\{\mu_{x0} \exp\left(\alpha_{xy}\right), \mu_{0y} \exp\left(\gamma_{xy}\right)\right\} = m_y. \end{cases}$$

Defining p with $\mu_{x0} = e^{p_x}$ and $\mu_{0y} = e^{-p_y}$, and Q with

$$\begin{cases} Q_x(p) = e^{p_x} + \sum_y \min\left\{e^{p_x + \alpha_{xy}}, e^{\gamma_{xy} - p_y}\right\} - n_x \\ Q_y(p) = -e^{-p_y} - \sum_x \min\left\{e^{p_x + \alpha_{xy}}, e^{\gamma_{xy} - p_y}\right\} + m_y, \end{cases}$$

the system rewrites as $Q(p) = 0$, and one can check that Q is a continuous Z-function with strictly isotone aggregates (hence an M-function). The general theory applies: there exist both a sub- and a supersolution (obtained respectively by setting all prices very low or very high), and therefore the equilibrium exists and is unique.

In the full assignment case with $\sum_x n_x = \sum_y m_y$ and no outside option, we normalize $p_{y_0} = \pi$ and the problem becomes

$$\begin{cases} Q_x(p) = \sum_y \min\left\{e^{p_x+\alpha_{xy}}, e^{\gamma_{xy}-p_y}\right\} - n_x, & x \in X \\ Q_y(p) = -\sum_x \min\left\{e^{p_x+\alpha_{xy}}, e^{\gamma_{xy}-p_y}\right\} + m_y, & y \in Y\backslash\{y_0\}. \end{cases}$$

Here $Q : \mathbb{R}^{X\cup Y\backslash\{y_0\}} \to \mathbb{R}^{X\cup Y\backslash\{y_0\}}$ is a Z-function with isotone aggregates (hence an M_0-function). However, because of the min's, Q may not be responsive for some values of p, and for that reason it is unclear whether the Jacobi sequence is well-defined or not in this case. Finding conditions under which the Jacobi algorithm works in the full assignment case with non-transferable utility thus remains an open problem for now.

4.6 References and Notes

This section draws upon the book by Roth and Sotomayor [8]. It also builds on the papers by Galichon, Kominers and Weber [5], Adachi [9] and Gale and Shapley [7]. Additional references include the working papers by Galichon and Léger [3], and Galichon, Hsieh and Sylvestre [11].

Acknowledgements Both authors gratefully acknowledge support from European Research Council consolidator grant EQUIPRICE, ERC-CoG No. 866274.

References

1. J. Ortega, W. Rheinboldt, *Iterative Solution of Nonlinear Equations in Several Variables*. Classics in Applied Mathematics (Society for Industrial and Applied Mathematics, 1970)
2. A. Galichon, L. Samuelson, L. Vernet, Theor. Econ. **20**, 941 (2025)
3. A. Galichon, F. Léger, Z-mappings (2023)
4. S. Berry, A. Gandhi, P. Haile, Econometrica **81**(5), 2087 (2013)
5. A. Galichon, S.D. Kominers, S. Weber, J. Polit. Econ. **127**(6), 2875 (2019)
6. L. Chen, E. Choo, A. Galichon, S. Weber, Existence and uniqueness in matching function equilibria with full assignment (2021)
7. D. Gale, L.S. Shapley, Am. Math. Mon. **69**(1), 9 (1962)
8. A. Roth, M. Sotomayor, *Two-Sided Matching: A Study in Game-Theoretic Modeling and Analysis* (Cambridge University Press, Cambridge, 1990)
9. H. Adachi, Econ. Lett. **68**(1), 43 (2000)
10. D.E. Knuth, *Stable Marriage and Its Relation to Other Combinatorial Problems: An Introduction to the Mathematical Analysis of Algorithms* (American Mathematical Society, 1997)
11. A. Galichon, Y.W. Hsieh, M. Sylvestre, Monotone comparative statics for submodular functions, with an application to aggregated deferred acceptance (2023)

The Harmonic Approximation to Quadratic Optimal Transport

Lukas Koch and Felix Otto

1 A Brief Introduction

These notes grew out of a couple of lecture series given by the second author at 2022 summer schools on the topic of optimal transportation, its regularity theory, and its application to the matching of random point clouds, see for instance https://kantorovich.org/event/2022-optimal-transport-summer-school/schedule/. These lectures presented results in [7], [6], and [8].

The traditional approach to regularity [3] and partial regularity [4, 5] for optimal transportation relies on the seminal regularity theory [2] for the corresponding Euler-Lagrange equation, the Monge-Ampère equation, based on the comparison principle. The variational approach introduced in [7] avoids these arguments and was first used to re-derive the partial regularity result of [5], and then in [9] to re-derive [4] for more general cost functions of quadratic behavior.

The added value of the variational approach lies in its robustness, in particular in its ability to deal with general measures: There is no need to have a Lebesgue density bounded away from zero and infinity in order to allow for explicit barrier functions as in the approach based on comparison principle. This robustness for instance allows to give a mesoscopic characterization of the optimal transport between what is allowed to be an atomic measure and the uniform distribution [6, Corollary 1.1], and allows to analyze the matching between independent copies of the Poisson point process [8]. We refer to www.mis.mpg.de/services/media/imprs-ringvorlesung-2022.html for a gentle introduction into this aspect of matching.

L. Koch (✉)
Department of Mathematics, University of Sussex, Falmer Campus, UK
e-mail: lukas.koch@sussex.ac.uk

F. Otto
Max Planck Institute for Mathematics in the Sciences, Leipzig, Germany
e-mail: felix.otto@mis.mpg.de

© The Author(s), under exclusive license to Springer Nature Switzerland AG 2026

Y. Kim et al. (eds.), *The Mathematics of Monge-Kantorovich Optimal Transport*,
Springer Proceedings in Mathematics & Statistics 552,
https://doi.org/10.1007/978-3-032-23146-8_2

In analogy to de Giorgi's strategy for ε-regularity of minimal surfaces, the core of the variational regularity theory is a harmonic approximation result [6, Theorem 1.4], which we will focus on in these notes. Hence we do not discuss the literature further, but refer to [9] for a careful review of the literature, and the connection to minimal surface theory. Compared to [6, Section 3], which we mainly rely on, these notes come with more motivations, less PDE machinery, and a couple of simplifications. They allow for an independent reading.

1.1 Standing Assumptions and Language

Throughout the entire text, we will consider two non-negative (finite) measures λ and μ on $\mathbb{R}^d$ with $\lambda(\mathbb{R}^d) = \mu(\mathbb{R}^d)$, which one should think of as two different spatial distributions of the same amount of mass. It is convenient to assume that λ and μ have bounded support. A non-negative measure π on the product space $\mathbb{R}^d \times \mathbb{R}^d$ is called admissible if its marginals are given by λ and μ, which spelled out means

$$\int \zeta(x)\pi(dxdy) = \int \zeta d\lambda \quad \text{and} \quad \int \zeta(y)\pi(dxdy) = \int \zeta d\mu \tag{1}$$

for all continuous and compactly supported functions ("test functions") ζ on $\mathbb{R}^d$. One should think of π as one possible way of transporting the mass as distributed according to λ into the shape as described by μ (a "transport/transference plan"). An admissible π is called optimal if it minimizes

$$\int |x-y|^2 d\pi \overset{\text{short for}}{=} \int_{\mathbb{R}^d \times \mathbb{R}^d} |x-y|^2 \pi(dxdy),$$

which one interprets as a total transportation cost, since it integrates the cost of transporting a unit mass from x to y, which here is given by the *square* of the Euclidean distance. This is Kantorovich' relaxation of Monge's problem, applied to the quadratic cost function. The infimum (which actually is attained)

$$W^2(\lambda, \mu) = \inf\{ \int |x-y|^2 d\pi \mid \pi \text{ is admissible for } \lambda, \mu \} \tag{2}$$

defines a distance function W, which we call Wasserstein distance.

2 Connection of Optimal Transportation and the Neumann Problem for the Poisson Equation

In this section, we motivate the connection between optimal transportation (OT) and the Neumann boundary value problem for the Poisson equation.

2.1 *Trajectories*

For the above connection, it is convenient to adopt a dynamical view upon OT, identifying a pair (x, y) of (matched) points with the (straight) trajectory

$$[0, 1] \ni t \mapsto X(t) := ty + (1-t)x. \tag{3}$$

Given an optimal transfer plan π for λ, μ, we ask the question on how to choose a function ϕ in such a way that its gradient $\nabla\phi$ captures the velocity of the trajectories, meaning

$$\dot{X}(t) \approx \nabla\phi(X(t)) \quad \text{for } (x, y) \in \operatorname{supp}\pi. \tag{4}$$

As we shall see, the answer relates to the Poisson equation $-\Delta\phi = \mu - \lambda$.

We are interested in connecting to a boundary value problem for the Poisson equation on some domain, say a ball B_R of some radius R (to be optimized later) and center w. l. o. g. given by the origin. We are thus led to restrict ourselves[1] to the set of trajectories that spend some time in the closure $\bar{B}_R$:

$$\Omega := \{ (x, y) \mid \exists t \in [0, 1] \; X(t) \in \bar{B}_R \}. \tag{5}$$

To every $(x, y) \in \Omega$, we associate the entering and exiting times $0 \le \sigma \le \tau \le 1$ of the corresponding trajectory

$$\begin{aligned} \sigma &:= \min\{t \in [0, 1] \mid X(t) \in \bar{B}_R\}, \\ \tau &:= \max\{t \in [0, 1] \mid X(t) \in \bar{B}_R\}, \end{aligned} \tag{6}$$

see also Fig. 1. (Note that some trajectories may both enter and exit.)

Given a transfer plan π, we keep track of *where* the trajectories enter and exit B_R, which is captured by two (non-negative) measures f and g concentrated on ∂B_R, defined through

[1] we will proceed to a further restriction in (12)

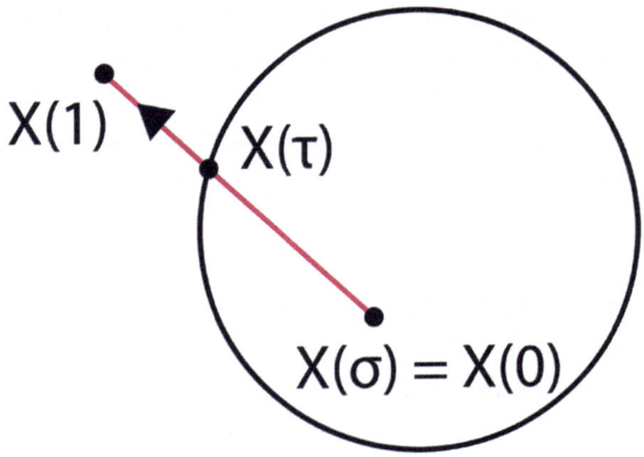

Fig. 1 Entering and exiting times of trajectories

$$\int \zeta df = \int_{\Omega\cap\{X(\sigma)\in\partial B_R\}} \zeta(X(\sigma))d\pi, \tag{7}$$

$$\int \zeta dg = \int_{\Omega\cap\{X(\tau)\in\partial B_R\}} \zeta(X(\tau))d\pi \tag{8}$$

for all test functions functions ζ. Note that the set of trajectories $\Omega\cap\{X(\sigma)\in\partial B_R\}$ implicitly defines a Borel measurable subset of $\mathbb{R}^d\times\mathbb{R}^d$, namely the pre-image under the mapping (3), which is continuous from $\mathbb{R}^d\times\mathbb{R}^d$ into $C^0([0,1])$. Hence the integration against π in (7) is legitimate.

Lemma 1 *We have for any admissible π and any continuously differentiable function ϕ on $\bar{B}_R$*

$$\begin{aligned}&\int_\Omega\int_\sigma^\tau |\dot{X}(t)-\nabla\phi(X(t))|^2 dt d\pi\\ &=\int_\Omega\int_\sigma^\tau |\dot{X}(t)|^2 dt d\pi + \int_\Omega\int_\sigma^\tau |\nabla\phi(X(t))|^2 dt d\pi\\ &-2\int_{B_R}\phi d(\mu-\lambda) - 2\int_{\partial B_R}\phi d(g-f).\end{aligned} \tag{9}$$

For later purpose, we record

$$\lambda(B_R)+f(\partial B_R)=\mu(B_R)+g(\partial B_R). \tag{10}$$

Proof of Lemma 1. For identity (9) we note that for the mixed term we have by the chain rule $\dot{X}(t)\cdot\nabla\phi(X(t))=\frac{d}{dt}[\phi(X(t))]$ and thus by the fundamental theorem of calculus $\int_\sigma^\tau \dot{X}(t)\cdot\nabla\phi(X(t))dt=\phi(X(\tau))-\phi(X(\sigma))$. In view of definition (6) we either have $X(\sigma)\in\partial B_R$ or $X(\sigma)\in B_R$. By definition (6) of σ the latter implies $\sigma=0$ and thus $X(\sigma)=x$, so that the constraint $(x,y)\in\Omega$ may be dropped. Hence $\int_\Omega\phi(X(\sigma))d\pi=\int_{\Omega\cap\{X(\sigma)\in\partial B_R\}}\phi(X(\sigma))d\pi+\int_{\{x\in B_R\}}\phi(x)d\pi$. By definition (7), the first integral is $\int \phi df$. By admissibility (1) of π, the second integral is $\int_{B_R} \phi d\lambda$. Likewise, one obtains $\int_\Omega \phi(X(\tau))d\pi = \int \phi dg + \int_{B_R} \phi d\mu$.

Specifying to $\phi = 1$, and thus $\nabla\phi = 0$ so that the mixed term vanishes, we learn (10) from the above two identities.

2.2 *Perturbative Regime*

We will focus on a "perturbative regime", which comes in form of two local smallness conditions. Any smallness condition has to be formulated in a non-dimensionalized way, which we implement by expressing this local smallness condition on a ball of non-dimensionalized radius, it will be convenient to take 5 as this radius.

The first smallness condition involves the data (thus the letter D), that is, the two measures λ and μ. We monitor how close these measures are to the Lebesgue measure on B_5. It is natural to quantify this in terms of the Wasserstein distance, see (2). Since the mass $\lambda(B_5)$ in general is not equal to the Lebesgue volume $|B_5|$, we have to split this into two: We monitor how Wasserstein-close the restriction $\lambda\llcorner B_5$ is to the uniform measure $\kappa_\lambda dx\llcorner B_5$, where $\kappa_\lambda := \frac{\lambda(B_5)}{|B_5|}$, and we monitor how close this density κ_λ is to unity. It is convenient to do both on the squared level:

$$\begin{aligned} D := {} & W^2(\lambda\llcorner B_5, \kappa_\lambda dx\llcorner B_5) + (\kappa_\lambda - 1)^2 \\ & + \text{same expression with } \lambda \rightsquigarrow \mu. \end{aligned} \tag{11}$$

In view of the localization (11), it is convenient to further restrict the set of trajectories, imposing that they start or end in B_4, thereby replacing (5) by

$$\Omega = \{(x, y) \in (B_4 \times \mathbb{R}^d) \cup (\mathbb{R}^d \times B_4) |\, \exists t \in [0, 1]\ X(t) \in \bar{B}_R\}. \tag{12}$$

The second smallness condition involves the solution itself, i. e. π. It monitors the length of trajectories that start or end in B_5. It does so in a square-averaged sense, like the total cost function itself. In fact, it is a localization of the cost functional (or energy, thus the letter E):

$$E := \int_{(B_5\times\mathbb{R}^d)\cup(\mathbb{R}^d\times B_5)} |x - y|^2 d\pi. \tag{13}$$

We expect (and shall rigorously argue in Sect. 3.4) that in the perturbative regime $E + D \ll 1$[2] and for a suitable $R \in [2, 3]$ we have for the second r. h. s. term in (9)

$$\int_\Omega \int_\sigma^\tau |\nabla\phi(X(t))|^2 dt d\pi \approx \int_{B_R} |\nabla\phi|^2. \tag{14}$$

[2] Here and in the following we use the notation $\ll 1$ to mean that there is $\varepsilon > 0$ such that the statement holds if $(E + D) \le \varepsilon$.

Indeed, for $E \ll 1$, trajectories are short so that

$$\int_\Omega \int_\sigma^\tau |\nabla\phi(X(t))|^2 dt d\pi \approx \int_{\{x \in B_R\}} |\nabla\phi(x)|^2 d\pi = \int_{B_R} |\nabla\phi|^2 d\lambda,$$

where the last identity follows from admissibility (1). Furthermore, for $D \ll 1$, λ is close to Lebesgue so that

$$\int_{B_R} |\nabla\phi|^2 d\lambda \approx \int_{B_R} |\nabla\phi|^2.$$

2.3 Connection to the Neumann Problem for the Poisson Equation

Hence in order to achieve (4), in view of (9) and (14), we are led to minimize

$$\int_{B_R} |\nabla\phi|^2 - 2\int_{B_R} \phi d(\mu - \lambda) - 2\int_{\partial B_R} \phi d(g - f) \tag{15}$$

in ϕ. A minimizer ϕ of (15), if it exists as a continuously differentiable function on $\bar{B}_R$, would be characterized by the Euler-Lagrange equation

$$\int_{B_R} \nabla\zeta \cdot \nabla\phi - \int_{B_R} \zeta d(\mu - \lambda) - \int_{\partial B_R} \zeta d(g - f) = 0 \tag{16}$$

for all continuously differentiable test functions ζ on B_R. If ϕ even exists as a twice continuously differentiable function on $\bar{B}_R$, we could appeal to the calculus identity $\nabla\zeta \cdot \nabla\phi = \nabla \cdot (\zeta\nabla\phi) - \zeta\Delta\phi$ and the divergence theorem in form of $\int_{B_R} \nabla \cdot (\zeta\nabla\phi) = \int_{\partial B_R} \zeta\nu \cdot \nabla\phi$, where $\nu(x) = \frac{x}{R}$ denotes the outer normal to ∂B_R in a point x, to obtain the integration by parts formula

$$\int_{B_R} \nabla\zeta \cdot \nabla\phi = \int_{B_R} \zeta(-\Delta\phi) + \int_{\partial B_R} \zeta\nu \cdot \nabla\phi. \tag{17}$$

Hence (16) can be reformulated and regrouped as

$$\int_{B_R} \zeta(-\Delta\phi - d(\mu - \lambda)) + \int_{\partial B_R} \zeta(\nu \cdot \nabla\phi - d(g - f)) = 0. \tag{18}$$

Considering first all test functions ζ's that vanish on ∂B_R, we learn from (18) that $-\Delta\phi = \mu - \lambda$ distributionally in B_R. Since $\mu - \lambda$ is a bounded measure, the first term in (18) thus vanishes also for test functions that do not vanish on ∂B_R. Hence the second term in (18) vanishes individually, which means $\nu \cdot \nabla\phi = g - f$ distributionally on ∂B_R. Hence we end up with what is called the Poisson equation with Neumann boundary conditions

$$-\Delta\phi = \mu - \lambda \text{ in } B_R, \quad \nu \cdot \nabla\phi = g - f \text{ on } \partial B_R. \tag{19}$$

This is a classical elliptic boundary value problem, which for sufficiently regular $\mu - \lambda$ and $g - f$ has a unique twice differentiable solution, provided (10) holds, and

$$\int_{B_R} \phi = 0 \tag{20}$$

is imposed. This motivates the connection between optimal transportation and the (short) Neumann-Poisson problem.

However, for rough (like sum of Diracs) measures λ, μ, and thus also rough measures f, g, the solution ϕ of (19), even if it exists for this linear problem, will be rough, too. In particular, (14) may not be true; even worse, both the l. h. s. and the r. h. s. might be infinite. Hence we shall approximate both $\mu - \lambda$ and $g - f$ by smooth functions (in fact, we shall approximate $\mu - \lambda$ by a constant function). The best way to organize the output of Lemma 1 is given by

Corollary 1 *We have for any admissible π and any twice continuously differentiable function ϕ on $\bar{B}_R$*

$$\begin{aligned}
&\int_\Omega \int_\sigma^\tau |\dot{X}(t) - \nabla\phi(X(t))|^2 dt d\pi \\
&\leq \int_\Omega |x - y|^2 d\pi - \int_{B_R} |\nabla\phi|^2 \\
&+ 2\int_{B_R} \phi(-\Delta\phi - d(\mu - \lambda)) + 2\int_{\partial B_R} \phi(\nu \cdot \nabla\phi - d(g - f)) \\
&+ \int_\Omega \int_\sigma^\tau |\nabla\phi(X(t))|^2 dt d\pi - \int_{B_R} |\nabla\phi|^2.
\end{aligned} \tag{21}$$

As we argued, see (14), we expect the term in last line (21) to be of higher order. The integrals in the second r. h. s. line can be made small by approximately solving (19) – and there will be a trade-off between making the last line and the second line small. However, the main open task is to argue, based on the optimality of π, that the difference in the first r. h. s. line is small for an approximate solution of (19). Hence we turn to this task before dealing with the second and third line in Sect. 3.4.

PROOF OF COROLLARY 1. The upgrade of identity (9) to inequality (21) relies on

$$\int_\Omega \int_\sigma^\tau |\dot{X}(t)|^2 dt d\pi \leq \int_\Omega |x - y|^2 d\pi, \tag{22}$$

$$\int_{B_R} |\nabla\phi|^2 = \int_{B_R} \phi(-\Delta\phi) + \int_{\partial B_R} \phi\nu \cdot \nabla\phi, \tag{23}$$

after adding and subtracting $2\int_{B_R}|\nabla\phi|^2$. Inequality (22) follows from $\int_\sigma^\tau|\dot X(t)|^2dt \le \int_0^1|\dot X(t)|^2dt = |x-y|^2$. Identity (23) follows from (17) for $\zeta=\phi$.

2.4 Localizing Optimality

As mentioned after Corollary 1, the main open task is to estimate the first r. h. s. line of (21). For this, we will (for the first time) use that π is optimal. In order to connect to the Neumann-Poisson problem on B_R, we need to leverage optimality in a localized way. Of course, it will in general not be true that the cost of π localized to $(B_R\times\mathbb{R}^d)\cup(\mathbb{R}^d\times B_R)$ is estimated by the transportation cost between the localized measures $\lambda\llcorner B_R$ and $\mu\llcorner B_R$. However, this is almost true if one adds the distribution of the entering points f, see (7), and exiting points g, see (8), respectively:

Lemma 2 *For π optimal we have*

$$\Big(\int_\Omega|x-y|^2d\pi\Big)^{\frac12}\le W(\lambda\llcorner B_R+f,\mu\llcorner B_R+g)$$
$$+\Big(2\int_{\Omega\cap\{\exists t\in[0,1]\ X(t)\in\partial B_R\}}|x-y|^2d\pi\Big)^{\frac12}. \qquad (24)$$

Lemma 2 controls the transportation cost coming from those trajectories that spend some time in $\bar B_R$, which amounts to the l. h. s. of (24) according to definition (12), by an OT problem localized to $\bar B_R$ as described by the first r. h. s. term. It does so up to the transportation cost coming from those (fewer) trajectories that cross (or touch) the boundary ∂B_R, see the second r. h. s. term. We shall argue in Lemma 5, cf. (66), that this last term (without the square root) is $o(E)$ for a good choice of R.

As its form suggests, (24) has the structure of a triangle inequality. In fact, its proof has similarities with the proof of the triangle inequality for W, using a disintegration (or conditioning) argument, c. f. [11, Section 5.1].

Proof of Lemma 2. We now introduce the distribution of $x=X(0)$ under π conditioned on the event that the trajectory X enters at $z\in\partial B_R$. In less probabilistic and more measure-theoretic terms ("dis-integration"), we introduce the (weakly continuous) family of probability measures $\{\lambda_z\}_{z\in\partial B_R}$ such that

$$\int_{\Omega\cap\{X(\sigma)\in\partial B_R\}}\zeta(x,X(\sigma))\pi(dxdy)=\int_{\partial B_R}\int\zeta(x,z)\lambda_z(dx)f(dz), \qquad (25)$$

which is possible by (7). Here, ζ is an arbitrary test function on $\mathbb{R}^d\times\mathbb{R}^d$. Likewise, we introduce the probability distribution $\{\mu_w\}_{w\in\partial B_R}$ of the end points of trajectories that exit in w:

$$\int_{\Omega\cap\{X(\tau)\in\partial B_R\}}\zeta(X(\tau),y)\pi(dxdy)=\int_{\partial B_R}\int\zeta(w,y)\mu_w(dy)g(dw). \qquad (26)$$

Fig. 2 The terms in (27)

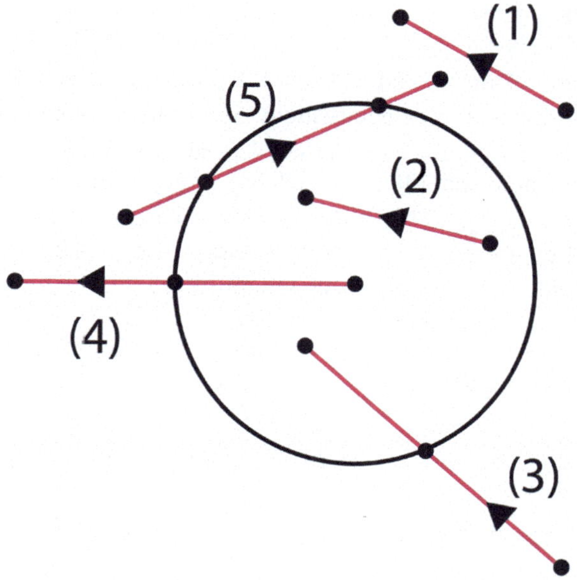

Let $\bar{\pi}$ denote an optimal plan for $W(\lambda \llcorner B_R + f, \mu \llcorner B_R + g)$. Equipped with these objects, we now define a competitor $\tilde{\pi}$ for π that mixes π with $\bar{\pi}$, in the sense that it takes the trajectories from π that stay outside of $\bar{B}_R$, the trajectories from $\bar{\pi}$ that stay inside (the open) B_R, and concatenates trajectories X from π that enter or exit $\bar{B}_R$ with trajectories of $\bar{\pi}$ that start or end in ∂B_R:

$$
\begin{aligned}
\int \zeta(x,y)\tilde{\pi}(dxdy) = &\int_{\Omega^c} \zeta(x,y)\pi(dxdy) \\
&+ \int_{B_R \times B_R} \zeta(x,y)\bar{\pi}(dxdy) \\
&+ \int_{\partial B_R \times B_R} \int \zeta(x,y)\lambda_z(dx)\bar{\pi}(dzdy) \\
&+ \int_{B_R \times \partial B_R} \int \zeta(x,y)\mu_w(dy)\bar{\pi}(dxdw) \\
&+ \int_{\partial B_R \times \partial B_R} \int \int \zeta(x,y)\mu_w(dy)\lambda_z(dx)\bar{\pi}(dzdw) \\
=: &\, (1) + (2) + (3) + (4) + (5),
\end{aligned}
\tag{27}
$$

see Fig. 2.

It is straightforward to see that $\tilde{\pi}$ has marginals λ and μ; by symmetry, it is sufficient to check the first condition in (1) by using (27) for a function $\zeta = \zeta(x)$: Since μ_w is a probability measure to the effect of $\int \zeta(x)\mu_w(dy) = \zeta(x)$, the second and the fourth r. h. s. term of (27) combine to $\int_{B_R \times \mathbb{R}^d} \zeta(x)\bar{\pi}(dxdy)$ because $\bar{\pi}$, like $\mu \llcorner B_R + g$, is supported in $\bar{B}_R$ (in dy). Likewise, the third and the fifth term combine to $\int_{\partial B_R \times \mathbb{R}^d} \int \zeta(x)\, \lambda_z(dx)\, \bar{\pi}(dzdy)$. By admissibility of $\bar{\pi}$, the combination of the second and fourth term gives $\int_{B_R} \zeta(x)\, \mu(dx)$, which as in the proof of Lemma 1 (by admissibility of π) can be seen to be $\int_{\Omega \cap \{X(\sigma) \in B_R\}} \zeta(x)\pi(dxdy)$. Since $\int \zeta(x)\lambda_z(dx)$ does not depend on y, for the same reason, the combination of the third and fifth term renders $\int_{\partial B_R} \zeta(z) f(dz)$, which by definition (7) is equal to $\int_{\Omega \cap \{X(\sigma) \in \partial B_R\}} \zeta(x)\, \pi(dxdy)$. Hence these four terms combine to $\int_\Omega \zeta(x)\pi(dxdy)$. Therefore, the r. h. s. of (27) collapses as desired to $\int \zeta(x)\pi(dxdy)$, which coincides with $\int \zeta(x)\lambda(dx)$ by admissibility of π.

By optimality of π, we have $\int |x-y|^2 d\pi \le \int |x-y|^2 d\bar{\pi}$; rewriting this as $\int_\Omega |x-y|^2 d\pi + \int_{\Omega^c} |x-y|^2 d\pi \le \int |x-y|^2 d\bar{\pi}$, and using (27) for $\zeta(x,y) = |x-y|^2$, we gather

$$\Big(\int_\Omega |x-y|^2 d\pi\Big)^{\frac{1}{2}} \le \|(f_2, f_3, f_4, f_5)\|, \tag{28}$$

where the four functions $f_2, \cdots, f_5 \ge 0$ are given by

$$f_2(x,y) := |x-y|, \quad f_5^2(z,w) := \int\int |x-y|^2 \mu_w(dy)\lambda_z(dx),$$
$$f_3^2(z,y) := \int |x-y|^2 \lambda_z(dx), \quad f_4^2(x,w) := \int |x-y|^2 \mu_w(dy),$$

and the vector-valued L^2-type norm is defined through

$$\begin{aligned}\|(f_2, f_3, f_4, f_5)\|^2 &= \int_{B_R \times B_R} f_2^2(x,y)\bar{\pi}(dxdy) + \int_{\partial B_R \times B_R} f_3^2(z,y)\bar{\pi}(dzdy) \\ &+ \int_{B_R \times \partial B_R} f_4^2(x,w)\bar{\pi}(dxdw) + \int_{\partial B_R \times \partial B_R} f_5^2(z,w)\bar{\pi}(dzdw). \end{aligned}\tag{29}$$

By the triangle inequality w. r. t. $L^2(\lambda_z)$ and $L^2(\mu_w)$, and using that λ_z, μ_w are probability measures, we obtain

$$f_3 \le |z-y| + \tilde{f}_3, \quad f_4 \le |x-w| + \tilde{f}_4, \quad f_5 \le |z-w| + \sqrt{2}\tilde{f}_5, \tag{30}$$

where the three functions $\tilde{f}_3, \tilde{f}_4, \tilde{f}_5 \ge 0$ are defined by

$$\tilde{f}_3^2(z,y) := \int |x-z|^2 \lambda_z(dx), \quad \tilde{f}_4^2(x,w) := \int |w-y|^2 \mu_w(dy), \tag{31}$$
$$\tilde{f}_5^2(z,w) := \tilde{f}_3^2(z,y) + \tilde{f}_4^2(x,w). \tag{32}$$

The factor of $\sqrt{2}$ in (30) arises because of $\tilde{f}_3 + \tilde{f}_4 \le \sqrt{2}\tilde{f}_5$.

From (30) we obtain by the triangle inequality for $\|\cdot\|$

$$
\begin{aligned}
\Big(\int_\Omega |x-y|^2 d\pi\Big)^{\frac{1}{2}} &\overset{(81)}{\leq} \|(f_2, f_3, f_4, f_5)\| \\
&\leq \|(|x-y|, |z-y|, |x-w|, |z-w|)\| + \sqrt{2}\|(0, \tilde{f}_3, \tilde{f}_4, \tilde{f}_5)\|,
\end{aligned} \tag{33}
$$

where we gave up a factor of $\sqrt{2}$ on $\tilde{f}_3, \tilde{f}_4$. By definition (29) of $\|\cdot\|$, the first r. h. s. term in (33) coincides with the square root of $\int_{\bar{B}_R\times\bar{B}_R}|x-y|^2\bar{\pi}$, which by optimality of $\bar{\pi}$ is $W(\lambda\llcorner B_R + f, \mu\llcorner B_R + g)$, as desired.

Using the definitions (29), (31), and (32), the square of the second r. h. s. term in (33) is equal to 2 times $\int_{\partial B_R\times\mathbb{R}^c}\int |x-z|^2\lambda_z(dx)\bar{\pi}(dzdy) + \int_{\mathbb{R}^d\times\partial B_R}\int |w-y|^2\mu_y(dy)\bar{\pi}(dxdw)$. Due to the admissibility of $\bar{\pi}$, this sum is equal to $\int_{\partial B_R}\int |x-z|^2\lambda_z(dx)f(dz) + \int_{\partial B_R}\int |w-y|^2\mu_y(dy)g(dw)$. By definitions (25) and (26), this coincides with $\int_{\Omega\cap\{X(\sigma)\in\partial B_R\}}|x-X(\sigma)|^2 d\pi + \int_{\Omega\cap\{X(\tau)\in\partial B_R\}}|X(\tau)-y|^2 d\pi$. Since we have $|x-X(\sigma)|^2+|X(\tau)-y|^2\leq|x-y|^2$, this sum is $\leq\int_{\Omega\cap(\{X(\sigma)\in\partial B_R\}\cup\{X(\tau)\in\partial B_R\})}|x-y|^2 d\pi$. Note that this set of integration coincides with $\Omega\cap\{\exists t\in[0,1]\ X(t)\in\partial B_R\}$, as desired.

2.5 *Constructing a Competitor Based on the Neumann-Poisson Problem*

As mentioned after Corollary 1, the remaining task is to estimate the first r. h. s. line of (21). For this, we will use Lemma 2 and construct a competitor for $W(\lambda\llcorner B_R + f, \mu\llcorner B_R + g)$ based on ϕ, the solution of the Neumann-Poisson problem (19), where we momentarily think of the measures λ, μ as having continuous densities with respect to the Lebesgue measure.

Lemma 3 *For ϕ as defined in (19),*

$$
W^2(\lambda\llcorner B_R + f, \mu\llcorner B_R + g) \leq \frac{1}{\min\{\min_{\bar{B}_R}\lambda, \min_{\bar{B}_R}\mu\}}\int_{B_R}|\nabla\phi|^2. \tag{34}
$$

Lemma 3 makes a second dilemma apparent: The intention was to use it in conjunction with Lemma 2 to obtain an estimate on the first r. h. s. line in (21). This however would require that we have $\min_{\bar{B}_R}\lambda, \min_{\bar{B}_R}\mu \gtrsim 1$, so a (one-sided) closeness of μ and λ to the Lebesgue measure in a strong topology, as opposed to the closeness in a weak topology as expressed by (11). Hence this provides another reason for approximating λ and μ by more regular versions.

Proof Lemma 3. The proof is short if one uses the Benamou-Brenier formulation in its distributional version, as we shall do. We recommend [11, Section 6.1] to the reader regarding more details on the Benamou-Brenier formulation. For every $t\in[0,1]$ we introduce the (singular non-negative) measure

$$\rho_t := t(\mu \llcorner B_R + g) + (1-t)(\lambda \llcorner B_R + f) \tag{35}$$

and the (t-independent) vector-valued measure

$$j_t := \nabla\phi dx \llcorner B_R. \tag{36}$$

We note that (19) in its distributional form of (16) can be re-expressed as

$$\frac{d}{dt}\int \zeta d\rho_t = \int \nabla\zeta \cdot dj_t \tag{37}$$

for all test functions ζ. In the jargon of the Benamou-Brenier formulation, which is inspired from continuum mechanics, ρ_t is a (mass) density, j_t is a flux, and (37) is the distributional version of the continuity equation $\partial_t\rho_t + \nabla \cdot j_t = 0$ expressing conservation of mass.

Following Benamou-Brenier one takes the Radon-Nikodym derivative $\frac{dj_t}{d\rho_t}$ of the (vectorial) measure j_t w. r. t. ρ_t (it plays the role of an Eulerian velocity field), and considers the expression that corresponds to the total kinetic energy:

$$\frac{1}{2}\int \left|\frac{dj_t}{d\rho_t}\right|^2 d\rho_t := \sup\left\{\int \xi \cdot dj_t - \int \frac{1}{2}|\xi|^2 d\rho_t\right\} \in [0,\infty], \tag{38}$$

where the supremum is taken over all continuous vector fields ξ with compact support. Benamou-Brenier (see [11, Section 5.4]) gives

$$W^2(\rho_0,\rho_1) \le \int_0^1 \int \left|\frac{dj_t}{d\rho_t}\right|^2 d\rho_t dt. \tag{39}$$

Since in our case, j_t is supported in (the open) B_R, see (36), in the r. h. s. of (38) we may restrict ourselves to ξ supported in B_R. For these ξ's, definition (35) yields $\int \xi \cdot dj_t - \int \frac{1}{2}|\xi|^2 d\rho_t = \int_{B_R} \big(\xi \cdot \nabla\phi - \frac{1}{2}|\xi|^2(t\mu + (1-t)\lambda)\big)$. By Young's inequality in form of $\xi \cdot \nabla\phi \le \frac{1}{2}(t\mu + (1-t)\lambda)|\xi|^2 + \frac{1}{2(t\mu+(1-t)\lambda)}|\nabla\phi|^2$ we thus obtain for the r. h. s. of (39)

$$\int \left|\frac{dj_t}{d\rho_t}\right|^2 d\rho_t \le \int_{B_R} \frac{|\nabla\phi|^2}{t\mu + (1-t)\lambda} \le \frac{1}{\min\{\min_{\bar B_R}\lambda, \min_{\bar B_R}\mu\}}\int_{B_R}|\nabla\phi|^2.$$

Since by definition (35), the l. h. s. of (39) coincides with the l. h. s. of (34), we are done.

3 Harmonic Approximation

The purpose of this section is to establish that the displacement in an optimal plan π can locally be approximated by a harmonic gradient $\nabla\phi$ (by which we mean that for each Cartesian direction $i = 1, \cdots, d$, the component $\partial_i\phi$ is harmonic, as a consequence of $-\Delta\phi = const$). This holds provided we are in the perturbative regime, see Sect. 2.2, where E and D are defined. More precisely, given any fraction $0 < \theta \ll 1$, there exists a threshold $\varepsilon > 0$ for $E + D$ so that below that threshold, the l. h. s. of (40) is only a fraction θ of E, plus a possibly large multiple of D.

Proposition 1 *For every $\theta > 0$, there exist $\varepsilon(d, \theta) > 0$ and $C(d, \theta) < \infty$ such that the following holds. Let π be optimal for λ, μ; provided $E + D \leq \varepsilon$, there exists a harmonic ϕ on B_2 such that*[3]

$$\int_{(B_1\times\mathbb{R}^d)\cup(\mathbb{R}^d\times B_1)} |(y-x) - \nabla\phi(x)|^2 d\pi \leq \theta E + CD, \tag{40}$$

$$\int_{B_1} |\nabla\phi|^2 \leq C(E + D). \tag{41}$$

(The proof actually reveals an explicit dependence of ε and C on θ.) We will obtain $\nabla\phi$ by solving the Neumann-Poisson problem

$$-\Delta\phi = \frac{\mu(B_R)}{|B_R|} - \frac{\lambda(B_R)}{|B_R|} \text{ in } B_R \quad \text{and} \quad \nu\cdot\nabla\phi = \bar{g} - \bar{f} \text{ on } \partial B_R, \tag{42}$$

where $\bar{f}, \bar{g}$ are suitable regular approximations of f, g, which are constructed in the nonlinear approximation Lemma 4, which also guides the choice of $R \in [2, 3]$. (In fact, in Sect. 3.4, we will replace $\bar{g} - \bar{f}$ by its mollification.) We note that $\bar{f}, \bar{g} \in L^2(\partial B_R)$ provides sufficient regularity: Indeed, according to (23) and (42) we have $\int_{B_R} |\nabla\phi|^2 = \int_{\partial B_R} \phi(\bar{g} - \bar{f})$, recalling the normalization $\int_{B_R} \phi = 0$. Applying Cauchy-Schwarz and then the Poincaré-trace estimate $\int_{\partial B_R} \phi^2 \leq C_P \int_{B_R} |\nabla\phi|^2$, we obtain

$$\int_{B_R} |\nabla\phi|^2 \leq C_P \int_{\partial B_R} (\bar{g} - \bar{f})^2. \tag{43}$$

In particular, (41) is a consequence of (51), for a suitable choice of $R \in [2, 3]$.

By an application of Lemma 3 to the setting of (42), we have

$$W^2\Big(\frac{\lambda(B_R)}{|B_R|} dx \llcorner B_R + \bar{f}, \frac{\mu(B_R)}{|B_R|} dx \llcorner B_R + \bar{g}\Big) \leq \frac{1}{\min\{\frac{\lambda(B_R)}{|B_R|}, \frac{\mu(B_R)}{|B_R|}\}} \int_{B_R} |\nabla\phi|^2. \tag{44}$$

---

[3] The integral is well-defined, since due to (65), $\mathrm{supp}(\pi) \cap (B_1 \times \mathbb{R}^d) \cup (\mathbb{R}^d \times B_1) \subset B_2 \times \mathbb{R}^d$.

Working with (42) instead of (19) creates the additional task of estimating the first r. h. s. term (24) of Lemma 2 by the l. h. s. of (44), which is conveniently done with help of the triangle inequality:

$$\begin{aligned}W(\lambda\llcorner B_R + f, \mu\llcorner B_R + g) \le\ & W(\frac{\lambda(B_R)}{|B_R|}dx\llcorner B_R + \bar{f}, \frac{\mu(B_R)}{|B_R|}dx\llcorner B_R + \bar{g})\\ &+ W(\lambda\llcorner B_R, \frac{\lambda(B_R)}{|B_R|}dx\llcorner B_R) + \text{same term with } \lambda \rightsquigarrow \mu\\ &+ W(f, \bar{f}) + \text{same term with } f \rightsquigarrow g. \end{aligned} \tag{45}$$

We now return to the first r. h. s. line in Corollary 1; in view of the elementary

$$\begin{aligned}&\int_\Omega |x-y|^2 d\pi - \int_{B_R} |\nabla\phi|^2\\ &\le 2\Big(\int_\Omega |x-y|^2 d\pi\Big)^{\frac{1}{2}}\Big(\Big(\int_\Omega |x-y|^2 d\pi\Big)^{\frac{1}{2}} - \Big(\int_{B_R} |\nabla\phi|^2\Big)^{\frac{1}{2}}\Big),\end{aligned} \tag{46}$$

and noting that by definitions (12) and (13), the first r. h. s. factor is estimated by E; by Young's inequality, it suffices to estimate the second factor. Combining (24), (44) and (45) we see that it is $\le$

$$\Big(\frac{1}{(\min\{\frac{\lambda(B_R)}{|B_R|}, \frac{\mu(B_R)}{|B_R|}\})^{\frac{1}{2}}} - 1\Big)\Big(\int_{B_R} |\nabla\phi|^2\Big)^{\frac{1}{2}} \tag{47}$$

plus

$$\begin{aligned}&\Big(2\int_{\Omega\cap\{\exists t\in[0,1]\ X(t)\in\partial B_R\}} |x-y|^2 d\pi\Big)^{\frac{1}{2}}\\ &+ W(\lambda\llcorner B_R, \frac{\lambda(B_R)}{|B_R|}dx\llcorner B_R) + \text{same term with } \lambda \rightsquigarrow \mu\\ &+ W(f, \bar{f}) + \text{same term with } f \rightsquigarrow g.\end{aligned} \tag{48}$$

We expect (and will show for a good radius $R \in [2, 3]$) that in the regime $D = o(1)$, the prefactor on the r. h. s. of (47) is $O(\sqrt{D})$, and that the second line in (48) is $O(\sqrt{D})$, as consistent with (40). These two technicalities are stated in Lemma 6. The main task is thus to control the last line in (48).

3.1 Approximating the Boundary Data

The main remaining task is to identify a good radius $R \in [2, 3]$ and to construct $\bar{f}$ and $\bar{g}$. Again, there is a trade-off/conflict of interest:

- On the one hand, the Neumann boundary data $\bar{g} - \bar{f}$ have to be sufficiently regular so that the solution ϕ of (42) is. In particular, we need (41) (with B_1 replaced by the larger B_R) to obtain that the error (47) is $o(E + D)$. Via (43), this is ensured by (51) in the upcoming Lemma 4. In fact, it even yields uniform integrability of $|\nabla\phi|^2$ on B_R, which is crucial to show that also the last line in (21) is $o(E + D)$.
- On the other hand, $(\bar{f}, \bar{g})$ has to be sufficiently close to (f, g). In particular, in view of the last term in (48) we need $W^2(f, \bar{f}) + W^2(g, \bar{g}) = o(E) + O(D)$. This is ensured by (50) in the upcoming Lemma 4. Here, as for (24), we will eventually need to appeal to $\int_1^2 \int_{\{\exists t\in[0,1]\ X(t)\in\partial B_R\}} |x - y|^2 d\pi\, dR = o(E + D)$, see Sect. 3.2.

In the upcoming approximation lemma we restrict to g for brevity.

Lemma 4 *We suppose that*

$$(X \in \Omega \implies y \in B_5) \ \ for\ (x, y) \in \mathrm{supp}\pi. \tag{49}$$

Then for every $R \in [2, 3]$ there exists a non-negative function $\bar{g}_R$ on ∂B_R such that

$$W^2(g_R, \bar{g}_R) \le 8\Big(\int_{\Omega\cap\{\exists t\in[0,1]\ X(t)\in\partial B_R\}} |x - y|^2 d\pi + D\Big), \tag{50}$$

$$\int_2^3 \int_{\partial B_R} \bar{g}_R^2\, dR \le 5^{d-1}\kappa_\mu(3E + D). \tag{51}$$

Note that we put an index R on g because the definition (8) obviously depends on R.

Proof of Lemma 4. We fix an $R \in [2, 3]$ and start with the construction of $\bar{g}_R$, momentarily returning to our short-hand notation $\bar{g}$. Let $\bar{\pi}$ be optimal for $W^2(\mu \llcorner B_5, \kappa_\mu dz \llcorner B_5)$; note that $\bar{\pi}$ is supported on $B_5 \times B_5$. We extend it (trivially) by the identity to $\mathbb{R}^d \times \mathbb{R}^d$; the extension (which we still call) $\bar{\pi}$ is admissible for $W^2(\mu, \kappa_\mu dz \llcorner B_5 + \mu \llcorner B_5^c)$. We retain

$$\int |y - z|^2 d\bar{\pi} = W^2(\mu \llcorner B_5, \kappa_\mu dz \llcorner B_5) \overset{(11)}{\le} D. \tag{52}$$

Like in the proof of the triangle inequality for the Wasserstein metric, we disintegrate $\bar{\pi}$ according to

$$\int \zeta(y, z)\bar{\pi}(dz|y)\mu(dy) = \int \zeta(y, z)\bar{\pi}(dydz), \tag{53}$$

since this family of (conditional) probability measures $\{\bar{\pi}(\cdot|y)\}_{y\in\mathbb{R}^d}$ allows us to define the measure $\tilde{\pi}$ on $\mathbb{R}^d \times \mathbb{R}^d \times \mathbb{R}^d$ via

$$\int \zeta(x, y, z)\tilde{\pi}(dxdydz) = \int\int \zeta(x, y, z)\bar{\pi}(dz|y)\pi(dxdy),$$

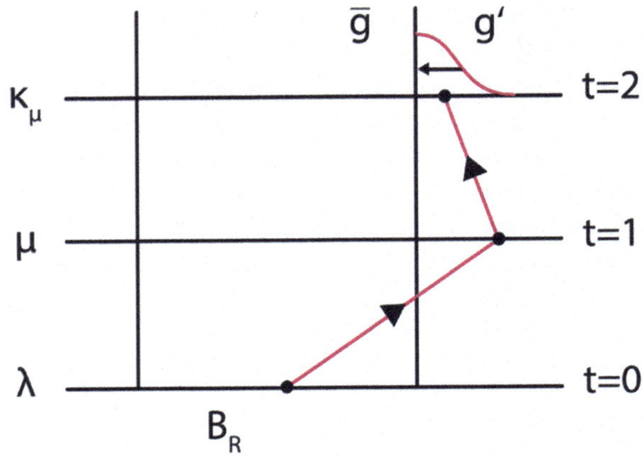

Fig. 3 Defining $\bar{g}$ and g'

which has the desired property

$$
\begin{aligned}
\text{marginal of } \tilde{\pi} \text{ w. r. t. } (x, y) &= \pi, \\
\text{marginal of } \tilde{\pi} \text{ w. r. t. } (y, z) &= \bar{\pi}.
\end{aligned} \tag{54}
$$

We still associate the trajectory X to a pair (x, y) as in (3). We extend this definition by associating to a triplet (x, y, z) the continuous piecewise affine trajectory defined by (3) followed by

$$X(t) = (t-1)z + (2-t)y \quad \text{for } t \in [1, 2].$$

We now are interested in the distribution g' of the endpoint $z = X(2)$ of those trajectories that exit $\bar{B}_R$ during the first time interval $[0, 1]$, i. e. those trajectories $X \in \Omega$ with $X(\tau) \in \partial B_R$, see Fig. 3. This distribution is defined by

$$\int \zeta dg' = \int_{\Omega \cap \{X(\tau) \in \partial B_R\}} \zeta(z) \tilde{\pi}(dxdydz). \tag{55}$$

We learn from (49) that $y = X(1) \in B_5$ for the trajectories X that contribute to (55). This yields $z = X(2) \in B_5$ by $\bar{\pi}(B_5, B_5^c) = 0$ according to our extension of $\bar{\pi}$. Hence we have

$$g' \text{ is supported in } B_5. \tag{56}$$

Therefore by the second item in (54) and the admissibility of $\bar{\pi}$ for $W^2(\mu, \kappa_\mu dz \llcorner B_5 + \mu \llcorner B_5^c)$, we obtain

$$\int \zeta dg' \le \int_{\{z \in B_5\}} \zeta(z) \bar{\pi}(dydz) = \kappa_\mu \int_{B_5} \zeta \quad \text{provided } \zeta \ge 0.$$

Hence g' admits a Lebesgue density, which we still denote by g' and that satisfies

$$g' \leq \kappa_\mu. \tag{57}$$

Finally, we radially project g' onto ∂B_R:

$$\int \zeta d\bar{g} = \int \zeta(R\frac{z}{|z|})g'(dz), \tag{58}$$

see Fig. 3. This concludes the construction of $\bar{g}$, we now turn to its estimate.

We start with (50) and note that an admissible plan for $W^2(g, \bar{g})$ is given by

$$\int_{\Omega\cap\{X(\tau)\in\partial B_R\}} \zeta(X(\tau), R\frac{z}{|z|})d\tilde{\pi}.$$

Indeed, on the one hand, for ζ only depending on the first variable, $\tilde{\pi}$ may be replaced by π according to the first item in (54) so that we obtain $\int \zeta dg$ by its definition (8). On the other hand, for ζ only depending on the second variable, we obtain $\int \zeta d\bar{g}$ by combining (55) and (58). Hence we have

$$W^2(g, \bar{g}) \leq \int_{\Omega\cap\{X(\tau)\in\partial B_R\}} |X(\tau) - R\frac{z}{|z|}|^2 d\tilde{\pi}. \tag{59}$$

Since $X(\tau) \in \partial B_R$, an elementary geometric argument on the radial projection yields for the integrand $|X(\tau) - R\frac{z}{|z|}| \leq 2|X(\tau) - z|$, so that $|X(\tau) - R\frac{z}{|z|}|^2 \leq 8(|x - y|^2 + |y - z|^2)$. This allows us to appeal to the compatibility (54) and to (52):

$$W^2(g, \bar{g}) \leq 8\Big(\int_{\Omega\cap\{\exists t\in[0,1]\ X(t)\in\partial B_R\}} |x - y|^2 d\pi + D\Big).$$

In preparation for establishing (51), we first provide an estimate of the measure g' defined in (55), which shows that it is concentrated near ∂B_R, see (60). By definition (55) we have

$$\int ||z| - R| dg' = \int_{\Omega\cap\{X(\tau)\in\partial B_R\}} ||z| - R|\tilde{\pi}(dxdydz).$$

Since $|X(\tau)| = R$, we may write $||z| - R| = ||z| - |X(\tau)|| \leq |x - y| + |y - z|$. Since $X(\tau) \in \partial B_R$ implies $\min_{[0,1]} |X| \leq R \leq \max_{[0,1]} |X|$, and $X(1) \in B_5$ by assumption (49), we thus obtain

$$\begin{aligned}&\int ||z| - R| dg' \\ &\leq \int_{\{y\in B_5\}\cap\{\min_{[0,1]}|X|\leq R\leq\max_{[0,1]}|X|\}} (|x - y| + |y - z|)\tilde{\pi}(dxdydz).\end{aligned}$$

Making the index R appear and integrating in R, this gives

$$\int_2^3 \int ||z| - R| dg'_R \, dR$$
$$\leq \int_{\{y\in B_5\}} (\max_{[0,1]} |X| - \min_{[0,1]} |X|)(|x-y| + |y-z|)\tilde{\pi}(dxdydz).$$

Using $\max_{[0,1]} |X| - \min_{[0,1]} |X| \leq |x-y|$ and then Young's inequality in form of $|x-y|(|x-y|+|y-z|) \leq \frac{3}{2}|x-y|^2 + \frac{1}{2}|y-z|^2$ we thus obtain from (54)

$$\int_2^3 \int ||z| - R| dg'_R \, dR$$
$$= \frac{3}{2} \int_{\{y\in B_5\}} |x-y|^2 \pi(dxdy) + \frac{1}{2} \int |y-z|^2 \tilde{\pi}(dydz).$$

By definition (13) and by (52) this turns into

$$\int_2^3 \int ||z| - R| dg'_R \, dR \leq \frac{3}{2} E + \frac{1}{2} D. \tag{60}$$

In order to pass from (60) to (51) we need

$$\int_{\partial B_R} \frac{1}{2} \bar{g}^2 \leq 5^{d-1} \kappa_\mu \int ||z| - R| dg'. \tag{61}$$

Here comes the argument for (61): By (57), it follows once we show for some density g' with (56)

$$\int_{\partial B_R} \frac{1}{2} \bar{g}^2 \leq 5^{d-1} (\operatorname{ess\,sup} g') \int ||z| - R| g' dz. \tag{62}$$

Introducing polar coordinates $z = r\hat{z}$ with $r \in (0, \infty)$ and $\hat{z} \in \partial B_1$, which are natural to re-express (58), (62) reduces to the single-variable statement

$$\frac{1}{2} \Big(\int g' r^{d-1} dr \Big)^2 \leq 5^{d-1} (\operatorname{ess\,sup} \kappa) \int |r - R| g' r^{d-1} dr.$$

It is convenient to rephrase this in terms of $\tilde{g} = g' r^{d-1}$; since because of (56) we have $\operatorname{ess\,sup} \tilde{g} \leq 5^{d-1} \operatorname{ess\,sup} g'$, it suffices to show for an arbitrary function $\tilde{g} \geq 0$ of $r \in (-\infty, \infty)$ that

$$\frac{1}{2} \Big(\int \tilde{g} dr \Big)^2 \leq (\operatorname{ess\,sup} \tilde{g}) \int |r - R| \tilde{g} dr. \tag{63}$$

The argument for (63) is elementary: By translation in r, we may assume $R = 0$; by homogeneity in $\tilde{g}$, we may assume $\operatorname{ess\,sup} \tilde{g} = 1$, that is, $\tilde{g} \in [0, 1]$. We now change perspective and seek to minimize the r. h. s. $\int |r|\tilde{g}dr$ under constraining the l. h. s. through prescribing $m = \int \tilde{g}dr$. By linearity of $\int |r|\tilde{g}dr$ in $\tilde{g}$, this functional assumes its minimum on extremal points w. r. t. the constraints $\tilde{g} \in [0, 1]$ and $\int \tilde{g}dr = m$. Those are characteristic functions of sets of Lebesgue measure m. Clearly, the set I with $|I| = m$ that minimizes $\int_I |r|dr$ is given by $I = [-\frac{m}{2}, \frac{m}{2}]$; the minimum is $\frac{m^2}{2}$, as desired.

3.2 Crossing Trajectories

In view of the second r. h. s. term in (24) of Lemma 2 and the first r. h. s. term in (50) of Lemma 4, we need to argue that for a suitable radius R, the trajectories crossing ∂B_R do not contribute much to E. This is the only part of the argument where we directly rely on the optimality criterion for π, namely the cyclical monotonicity of its support, c. f. [11, Section 1.6.2]. We just need it in form of plain monotonicity:

$$(x - x') \cdot (y - y') \geq 0 \quad \text{for all } (x, y), (x', y') \in \operatorname{supp}\pi. \tag{64}$$

This in fact implies that all trajectories are short in our regime of $E + D \ll 1$:

Lemma 5 *We have*

$$|x - y| = o(1) \quad \text{for } (x, y) \in \big((B_4 \times \mathbb{R}^d) \cup (\mathbb{R}^d \times B_4)\big) \cap \operatorname{supp}\pi, \tag{65}$$

$$\int_2^3 \int_{\Omega \cap \{\exists t \in [0,1]\ X(t) \in \partial B_R\}} |x - y|^2 d\pi dR = o(E + D), \tag{66}$$

$$\int_2^3 \pi\big(\Omega \cap \{\exists t \in [0, 1]\ X(t) \in \partial B_R\}\big) dR = o(1). \tag{67}$$

As a consequence of (66) we may indeed chose $R \in [2, 3]$ so that the terms in (24) and (50) are $o(E) + O(D)$. As a consequence of (65), also (49) is satisfied.

Proof of Lemma 5. We start by deriving (66) and (67) from (65). As in the proof of (51) we note that $X(t) \in \partial B_R$ implies $\min_{[0,1]} |X| \leq R \leq \max_{[0,1]} |X|$ and thus

$$\begin{aligned}
&\int_2^3 \int_{\Omega \cap \{\exists t \in [0,1]\ X(t) \in \partial B_R\}} |x - y|^2 \pi(dxdy) dR \\
&\leq \int_{(B_3 \times \mathbb{R}^d) \cup (\mathbb{R}^d \times B_3)} (\max_{[0,1]} |X| - \min_{[0,1]} |X|) |x - y|^2 \pi(dxdy).
\end{aligned}$$

Since $\max_{[0,1]} |X| - \min_{[0,1]} |X| \leq |x - y|$, (66) now follows from (65) by the definition (13). Likewise, we have

$$\int_2^3 \pi(\Omega \cap \{\exists t \in [0,1]\ X(t) \in \partial B_R\})dR$$
$$= o(1)\pi((B_3 \times \mathbb{R}^d) \cup (\mathbb{R}^d \times B_3)).$$

By (1) we obtain that $\pi((B_3 \times \mathbb{R}^d) \cup (\mathbb{R}^d \times B_3)) \le \lambda(B_5) + \mu(B_5)$, so that we may appeal to (11) to the effect of $\lambda(B_5) \le |B_5|(1+\sqrt{D}) \lesssim 1$. Here and in the following we use $\lesssim$ to mean up to constants that only depend on d.

We now turn to proving (65); a more explicit but pedestrian argument can be found in [6, Lemma 2.9]. Since $E \ll 1$, we expect there to be many short trajectories. We exploit monotonicity in order to upgrade this to a statement about all trajectories. By definition (12) and due to $(x \leftrightarrow y)$-symmetry, we may assume that $x \in B_4$. We will use (64) in form of

$$(x-y)\cdot(x-x') \le \frac{3}{2}|x-x'|^2 + \frac{1}{2}|x'-y'|^2. \tag{68}$$

For $\zeta \ge 0$ supported in B_5 we integrate (68) against $\zeta(x')\pi(dx'dy')$. Using the admissibility of π and the definition (13) of E, we find

$$(x-y)\cdot\int(x-x')\zeta(x')\lambda(dx')$$
$$\le \frac{3}{2}\int|x-x'|^2\zeta(x')\lambda(dx') + \frac{1}{2}(\sup\zeta)E. \tag{69}$$

By definition (11), $D \ll 1$ implies that $\lambda \llcorner B_5$ weakly close to $dx' \llcorner B_5$. In order to obtain a uniformity when applying this to (69), we now specify a class of functions ζ in (69): Fixing a smooth mask $\hat{\zeta} \ge 0$ supported in B_1 and some radius $0 < r \ll 1$ to be optimized later, we consider $\zeta(x') = \hat{\zeta}(\frac{x'-x_0'}{r})$ for any center x_0', only restricted by the requirement that ζ is supported in B_5. This class of functions is compact with respect to the uniform topology so that we have

$$\int(x-x')\zeta(x')\lambda(dx') = \int(x-x')\zeta(x')dx' + o_r(1),$$
$$\int|x-x'|^2\zeta(x')\lambda(dx') = \int|x-x'|^2\zeta(x')dx' + o_r(1)$$

as $D \ll 1$, uniformly in the center but at fixed radius (as the subscript in $o_r(1)$ is to indicate). Specifying $\hat{\zeta}$ to have unit integral and vanishing first moment, this yields

$$\int(x-x')\zeta(x')\lambda(dx') = r^d(x-x_0') + o_r(1),$$
$$\int|x-x'|^2\zeta(x')\lambda(dx') \lesssim r^d(|x-x_0'|^2 + r^2) + o_r(1).$$

This prompts the choice of the center $x_0' = x - r\frac{x-y}{|x-y|}$, which is admissible since in view of $x \in B_4$ we still have that ζ is supported in B_5, to the effect of

$$(x-y)\cdot\int(x-x')\zeta(x')\lambda(dx') = r^{d+1}|x-y| + o_r(|x-y|),$$
$$\int|x-x'|^2\zeta(x')\lambda(dx') \lesssim r^{d+2} + o_r(1).$$

Inserting this into (69) yields

$$|x-y| \lesssim r + \frac{E}{r^{d+1}} + o_r(|x-y|) + o_r(1).$$

We now may conclude: We first choose r so that the first r. h. s. term is small; we then choose E so that the second term is small, and we finally choose D so small such that the third term may be absorbed into the l. h. s. and such that the last term is small.

3.3 *Restricting the Data Term D*

While the data term D is defined w. r. t. to B_5, we rather need it w. r. t. B_R for our suitably chosen $R \in [2, 3]$. This is most prominent in the middle line of (48) and the prefactor of (47), but also related to the middle and last line on the r. h. s. of (21), as was discussed in Sect. 2.2. Annoyingly, this restriction property for D does not come for free and and requires arguments similar to the ones used in the proof of Lemma 4. By symmetry, it is enough to consider λ.

Lemma 6 *With* $\kappa_R = \frac{\lambda(B_R)}{|B_R|}$ *we have*

$$\int_2^3 W^2(\lambda\llcorner B_R, \kappa_R dx\llcorner B_R) + (\kappa_R - 1)^2 dR = O(D). \tag{70}$$

For the proof of Lemma 6, it is convenient to have the following two extensions of Lemma 3 on the relationship between OT and the Poisson-Neumann problem; the first is a rough generalization, the second provides the reverse relationship in a restricted setting.

Corollary 2 ([[1], Corollary 4.4.]) *For the* ϕ *defined in (19),*

$$W^2(\lambda\llcorner B_R + f, \mu\llcorner B_R + g) \le \frac{4}{\min_{\bar{B}_R}\mu}\int_{B_R}|\nabla\phi|^2. \tag{71}$$

Lemma 7 *Provided* $f, g \equiv 0$ *in (19),*

$$W^2(\lambda\llcorner B_R, \mu\llcorner B_R) \geq \frac{1}{\max\{\max_{\bar{B}_R}\lambda, \max_{\bar{B}_R}\mu\}} \int_{B_R} |\nabla\phi|^2. \tag{72}$$

Proof of Corollary 2. A proof based on the Benamou-Brenier formula was given in [1]. For the convenience of the reader, we provide an elementary proof here. We first argue that for arbitrary λ, μ and $0 \leq M < \infty$,

$$W(\lambda, \mu) \leq \frac{1}{\sqrt{1+M} - \sqrt{M}} W(\lambda + M\mu, (1+M)\mu). \tag{73}$$

Indeed, by scaling we have $W(\lambda, \mu) = \frac{1}{\sqrt{1+M}} W((1+M)\lambda, (1+M)\mu)$. By the triangle inequality, $W((1+M)\lambda, (1+M)\mu) \leq W((1+M)\lambda, \lambda + M\mu) + W(\lambda + M\mu, (1+M)\mu)$, which we combine with the obvious $W((1+M)\lambda, \lambda + M\mu) = W(M\lambda, M\mu)$. Once more by scaling, $W(M\lambda, M\mu) = \sqrt{M} W(\lambda, \mu)$, so that we may absorb.

We now argue that with help of Lemma 3 we obtain

$$W^2(\lambda\llcorner B_R + f, \mu\llcorner B_R + g) \leq \frac{1}{M(\sqrt{1+M} - \sqrt{M})^2} \frac{1}{\min_{\bar{B}_R}\mu} \int_{B_R} |\nabla\phi|^2,$$

which yields (71) under $M \uparrow \infty$. Indeed, we first apply (73) with (λ, μ) replaced by $(\lambda\llcorner B_R + f, \mu\llcorner B_R + g)$. We then use (34) to estimate $W^2(\lambda\llcorner B_R + f + M(\mu\llcorner B_R + g), (1+M)(\mu\llcorner B_R + g))$, noting that when taking the difference of $(1+M)\mu$ and $\lambda + M\mu$ and of $(1+M)g$ and $f + Mg$, the M-dependent terms drop out and thus does not affect the definition (19) of ϕ. It remains to observe that the minimum of the Lebesgue densities of $(1+M)\mu$ and $\lambda + M\mu$ on $\bar{B}_R$ is bounded below by $M\min_{\bar{B}_R}\mu$.

Proof of Lemma 7. We recall the Benamou-Brenier formulation from Lemma 3 and note that for the optimal (ρ_t, j_t)

$$W^2(\lambda\llcorner B_R, \mu\llcorner B_R) \geq \int_0^1 \int \left|\frac{dj_t}{d\rho_t}\right|^2 d\rho_t dt. \tag{74}$$

Setting $M := \max\{\max_{\bar{B}_R}\lambda, \max_{\bar{B}_R}\mu\} < \infty$, so that $\lambda\llcorner B_R, \mu\llcorner B_R \leq M dx\llcorner B_R$, we have by McCann's displacement convexity (in conjunction with the convexity of B_R), c.f. [11, Section 7.3],

$$\rho_t \leq M dx\llcorner B_R \tag{75}$$

for all $t \in [0, 1]$. In view of definition (38), we thus obtain for any test vector field ξ

$$\int \xi \cdot dj_t \leq \frac{1}{2}\int |\frac{dj_t}{d\rho_t}|^2 d\rho_t + \frac{M}{2}\int_{B_R} |\xi|^2.$$

After integration in $t \in [0, 1]$, replacing ξ by a $a\xi$ and optimizing in the constant $a > 0$, this yields $\int \xi \cdot dj_t \leq (\int |\frac{dj_t}{d\rho_t}|^2 d\rho_t \, M \int_{B_R} |\xi|^2)^{\frac{1}{2}}$. In conjunction with (74) this implies

$$\int_0^1 j_t dt \ll dx \llcorner B_R$$

$$\text{and} \quad \int_{B_R} |\int_0^1 j_t dt|^2 \leq M W^2(\lambda \llcorner B_R, \mu \llcorner B_R), \tag{76}$$

where we identify the measure $\int_0^1 j_t dt$ with its Lebesgue density.

From the admissibility condition (37) we obtain by integration in $t \in [0, 1]$ and using $\rho_{t=0} = \lambda \llcorner B_R$, $\rho_{t=1} = \mu \llcorner B_R$

$$-\nabla \cdot \int_0^1 j_t dt = \mu \llcorner B_R - \lambda \llcorner B_R \quad \text{distributionally on } \mathbb{R}^d.$$

Thus by (19) (with $f, g \equiv 0$), we have that $\int_0^1 j_t dt - \nabla\phi$ (with $\nabla\phi$ trivially extended beyond $\bar{B}_R$) is distributionally divergence-free in $\mathbb{R}^d$. Extending ϕ in a (compactly supported) C^1-manner outside of $\bar{B}_R$ and testing with this extension, because $\int_0^1 j_t dt$ is supported in B_R, cf. (76), we obtain $\int_{B_R} \nabla\phi \cdot (\int_0^1 j_t dt - \nabla\phi) = 0$. By Cauchy-Schwarz this yields

$$\int_{B_R} |\nabla\phi|^2 \leq \int_{B_R} |\int_0^1 j_t dt|^2.$$

Combining this with (76) we obtain (72).

Proof of Lemma 6. Inside this proof, we denote by π the optimal plan in $W^2(\lambda \llcorner B_5, \kappa_\lambda dx \llcorner B_5)$. Following the proof of Lemma 4, we monitor where entering and exiting trajectories end up:

$$\int \zeta df' = \int_{\Omega \cap \{X(0) \notin B_R\} \cap \{X(1) \in B_R\}} \zeta(X(1)) d\pi,$$
$$\int \zeta dg' = \int_{\Omega \cap \{X(0) \in B_R\} \cap \{X(1) \notin B_R\}} \zeta(X(1)) d\pi. \tag{77}$$

Clearly, the non-negative measures f', g' are supported in B_R and $\bar{B}_5 - B_R$, respectively, and satisfy $f', g' \leq \kappa_\lambda$. We introduce the corresponding mass densities w. r. t. B_R

$$\kappa_f := \frac{f'(\mathbb{R}^d)}{|B_R|} \leq \kappa_\lambda, \quad \kappa_g := \frac{g'(\mathbb{R}^d)}{|B_R|}. \tag{78}$$

We start with the triangle inequality

$$
\begin{aligned}
&W(\lambda\llcorner B_R, (\kappa_\lambda - \kappa_f + \kappa_g)dx\llcorner B_R) \\
&\le W(\lambda\llcorner B_R, \kappa_\lambda dx\llcorner B_R - f' + g') \\
&+ W(\kappa_\lambda dx\llcorner B_R - f' + g', (\kappa_\lambda - \kappa_f)dx\llcorner B_R + g') \\
&+ W((\kappa_\lambda - \kappa_f)dx\llcorner B_R + g', (\kappa_\lambda - \kappa_f + \kappa_g)dx\llcorner B_R).
\end{aligned} \tag{79}
$$

Restricting the optimal π to trajectories that start in B_R, we obtain an admissible plan for the first r. h. s. term of (79):

$$
W^2(\lambda\llcorner B_R, \kappa_\lambda dx\llcorner B_R - f' + g') \le W^2(\lambda\llcorner B_5, \kappa_\lambda dx\llcorner B_5) \le D.
$$

We now turn to the second r. h. s. term of (79), and let ϕ' denote the solution of

$$
-\Delta\phi' = f' - \kappa_f \text{ in } B_R, \quad \nu\cdot\nabla\phi' = 0 \text{ on } \partial B_R.
$$

On the one hand, we obtain from Corollary 2

$$
W(\kappa_\lambda dx\llcorner B_R - f' + g', (\kappa_\lambda - \kappa_f)dx\llcorner B_R + g') \le \frac{4}{\kappa_\lambda - \kappa_f}\int_{B_R}|\nabla\phi'|^2.
$$

On the other hand, we obtain from Lemma 7 and $f' \le \kappa_\lambda$

$$
\frac{1}{2\kappa_\lambda}\int_{B_R}|\nabla\phi'|^2 \le W(\kappa_\lambda dx\llcorner B_R, (\kappa_\lambda - \kappa_f)dx\llcorner B_R + f').
$$

The combination of these two inequalities yields

$$
\begin{aligned}
&W(\kappa_\lambda dx\llcorner B_R - f' + g', (\kappa_\lambda - \kappa_f)dx\llcorner B_R + g') \\
&\le \frac{8\kappa_\lambda}{\kappa_\lambda - \kappa_f} W(\kappa_\lambda dx\llcorner B_R, (\kappa_\lambda - \kappa_f)dx\llcorner B_R + f').
\end{aligned} \tag{80}
$$

In view of (80), we may treat the second r. h. s. term in (79) alongside of the last term, and focus on the latter. Like in Lemma 4, we introduce the projection $\bar{g}$ of g' onto ∂B_R, that is

$$
\int \zeta d\bar{g} = \int \zeta(\frac{Rx}{|x|})g'(dx). \tag{81}
$$

We start with the triangle inequality

$$
\begin{aligned}
&W((\kappa_\lambda - \kappa_f)dx\llcorner B_R + g', (\kappa_\lambda - \kappa_f + \kappa_g)dx\llcorner B_R) \\
&\le W((\kappa_\lambda - \kappa_f)dx\llcorner B_R + \bar{g}, (\kappa_\lambda - \kappa_f + \kappa_g)dx\llcorner B_R) + W(\bar{g}, g').
\end{aligned} \tag{82}
$$

According to Lemma 3, we have for the first term

$$W((\kappa_\lambda - \kappa_f)dx\llcorner B_R + \bar{g}, (\kappa_\lambda - \kappa_f + \kappa_g)dx\llcorner B_R) \leq \frac{1}{\kappa_\lambda - \kappa_f}\int_{B_R} |\nabla\bar{\phi}|^2, \quad (83)$$

where $\bar{\phi}$ is defined through

$$-\Delta\bar{\phi} = \kappa_g \text{ in } B_R, \quad \nu\cdot\nabla\bar{\phi} = \bar{g} \text{ on } \partial B_R.$$

We recall (61), which we may apply since $g' \leq \kappa_\lambda$ is supported in $\bar{B}_5 - B_R$, and which takes the form of

$$\int_{\partial B_R} \bar{g}^2 \leq 2\cdot 5^{d-1}\kappa_\lambda \int ||x| - R|dg'. \quad (84)$$

Combining this with (43), where ϕ is replaced by $\bar{\phi}$, we obtain

$$\int_{B_R} |\nabla\bar{\phi}|^2 \leq 2\cdot 5^{d-1}C_P\kappa_\lambda \int ||x| - R|dg'. \quad (85)$$

Before continuing with the r. h. s. of (85), we turn to the last term in (82). In view of (81), $\int \zeta(R\frac{x}{|x|}, x)g'(dx)$ defines an admissible plan, so that $W^2(\bar{g}, g') \leq \int |R\frac{x}{|x|} - x|^2 g'(dx)$. Noting that $|R\frac{x}{|x|} - x| = ||x| - R|$ and recalling that g' is supported in $\bar{B}_5 - B_R$, we obtain

$$W^2(\bar{g}, g') \leq 5\int ||x| - R|dg'.$$

In view of this and (85), we are lead to estimate $\int ||x| - R|dg'$. A simplification of the argument leading to (60) gives

$$\int_2^3 \int ||x| - R|dg'dR \leq D. \quad (86)$$

Since in view of (79) we have $\kappa_R = \kappa_\lambda - \kappa_f + \kappa_g$, in order to obtain (70), it remains to show $\kappa_f^2 + \kappa_g^2 \lesssim D$. Because of our assumption $D \ll 1$, this also deals with the pre-factors in (80) and (83). By symmetry, we may restrict to κ_g. By definitions (77), (78), and (81) we have $\kappa_g = \frac{1}{|B_R|}\bar{g}(\mathbb{R}^d)$, and thus by Cauchy-Schwarz $\kappa_g^2 \leq \frac{|\partial B_R|}{|B_R|^2}\int_{\partial B_R} \bar{g}^2$. Hence it remains to appeal to (84) and (86).

3.4 A Final Approximation and Proof of Proposition 1

Note that the two terms in (21) involving $\nabla\phi(X(t))$ make only sense for general π provided $\nabla\phi \in C^0(\bar{B}_R)$, which however is not ensured by $\bar{g} - \bar{f} \in L^2(\partial B_R)$ in (42).

Hence, a final – however more conventional – approximation argument is unavoidable: We approximate $\bar g - \bar f$ by its mollification $(\bar g - \bar f)_r$ on a scale $r > 0$, and denote by ϕ^r the corresponding solution of (42). With this replacement, (21) assumes the form

$$\int_\Omega \int_\sigma^\tau |\dot X(t) - \nabla\phi^r(X(t))|^2 dt d\pi \tag{87}$$

$$\leq \int_\Omega |x-y|^2 d\pi - \int_{B_R} |\nabla\phi^r|^2 \tag{88}$$

$$+2\int_{B_R} \phi^r \Big(\frac{\mu(B_R)}{|B_R|} - \frac{\lambda(B_R)}{|B_R|} - d(\mu-\lambda)\Big) \tag{89}$$

$$+2\int_{\partial B_R} \phi^r \big((\bar g - \bar f)_r - d(g-f)\big) \tag{90}$$

$$+\int_\Omega \int_\sigma^\tau |\nabla\phi^r(X(t))|^2 dt d\pi - \int_{B_R} |\nabla\phi^r|^2. \tag{91}$$

Before controlling the difference in line (91), for which the mollification was made, we address its effect on the lines (88) (where it is cumbersome), (89) (where it is beneficial), and (90) (where it is both).

We now fix the radius R such as to benefit from the Lemmas 4, 5, and 6. More precisely, taking the sum of the estimates (50) divided by $o(E) + O(D)$, (51) divided by $E + D$, (66) by $o(E + D)$, (67) by $o(1)$, and (70) by D,[4] we learn that there exists an $R \in [2, 3]$ and $\bar g, \bar f \in L^2(\partial B_R)$ with

$$W^2(f, \bar f) + W^2(g, \bar g) = o(E) + O(D), \tag{92}$$

$$\int_{\partial B_R} \bar f^2 + \bar g^2 = O(E + D), \tag{93}$$

$$\int_{\Omega \cap \{\exists t \in [0,1]\ X(t) \in \partial B_R\}} |x-y|^2 d\pi = o(E + D), \tag{94}$$

$$\pi(\Omega \cap \{\exists t \in [0, 1]\ X(t) \in \partial B_R\}) = o(1), \tag{95}$$

$$W^2(\lambda \llcorner B_R, \frac{\lambda(B_R)}{|B_R|} dx \llcorner B_R) + (\frac{\lambda(B_R)}{|B_R|} - 1)^2 = O(D), \tag{96}$$

---

[4] To pick out just the two terms of (50) and (67), what we mean is the following: According to these to statements, for any given $\delta > 0$ we have for sufficiently small $E + D$ that

$$\int_2^3 dR \frac{1}{\delta E + D} W^2(f_R, \bar f_R) + \frac{1}{\delta}\pi(\Omega \cap \{\exists t \in [0, 1]\ X(t) \in \partial B_R\}) \lesssim 1.$$

This means that there exists an $R \in [2, 3]$ such that the integrand is ≤ 1, which translates into

$$W^2(f_R, \bar f_R) \lesssim \delta E + D \quad \text{and} \quad \pi(\Omega \cap \{\exists t \in [0, 1]\ X(t) \in \partial B_R\}) \lesssim \delta,$$

which amounts to (92) and (95).

$$W^2(\mu \llcorner B_R, \frac{\mu(B_R)}{|B_R|} dx \llcorner B_R) + (\frac{\mu(B_R)}{|B_R|} - 1)^2 = O(D). \tag{97}$$

We begin with line (88): It follows from a standard mollification argument that the $\dot{H}^{-\frac{1}{2}}(\partial B_R)$-norm of $(\bar{g} - \bar{f})_r - (\bar{g} - \bar{f})$ is $O(r^{\frac{1}{2}})$ times the $L^2(\partial B_R)$-norm of $\bar{g} - \bar{f}$:

$$\int_{\partial B_R} \zeta\big((\bar{g} - \bar{f})_r - (\bar{g} - \bar{f})\big) \lesssim \Big(r \int_{B_R} |\nabla \zeta|^2 \int_{\partial B_R} (\bar{g} - \bar{f})^2\Big)^{\frac{1}{2}}. \tag{98}$$

Hence from (23) applied to (42), which yields that $\int_{B_R} |\nabla(\phi^r - \phi)|^2 = \int_{\partial B_R} (\phi^r - \phi)((\bar{g} - \bar{f})_r - (\bar{g} - \bar{f}))$, with the choice $\zeta = \phi^r - \phi$ in (98), we obtain

$$\int_{B_R} |\nabla(\phi^r - \phi)|^2 \lesssim r \int_{\partial B_R} (\bar{g} - \bar{f})^2. \tag{99}$$

Combining this with (43) yields

$$\int_{B_R} |\nabla \phi|^2 - \int_{B_R} |\nabla \phi^r|^2 \leq 2 \int_{B_R} \nabla(\phi - \phi^r) \cdot \nabla \phi \lesssim r^{\frac{1}{2}} \int_{\partial B_R} (\bar{g} - \bar{f})^2.$$

Hence in view of (93) we have

$$\int_{B_R} |\nabla \phi|^2 - \int_{B_R} |\nabla \phi^r|^2 \leq r^{\frac{1}{2}} O(E + D). \tag{100}$$

Equipped with (100) and the more obvious

$$\int_{B_R} |\nabla \phi|^2 = O(E + D), \tag{101}$$

we may now conclude the estimate of line (88). We refer back to Sect. 3, namely (46) followed by Young's inequality. Next to an $o(E)$ coming from the first factor in Young's inequality, this gives rise to the square of the term in (47) and the three terms in (48). According to (96) & (97) and (101), the former contribution is $O(D(E + D))$. According to (94), once more (96) & (97), and (92), the latter three contributions in (48) are $o(E + D)$, $O(D)$, and $o(E + D)$, respectively. Hence in combination with (100) we obtain

$$\begin{aligned} &\int_{\Omega} |x - y|^2 d\pi - \int_{B_R} |\nabla \phi|^2 \\ &\leq o(E) + O(D(E + D)) + o(E + D) + O(D) + r^{\frac{1}{2}} O(E + D), \end{aligned}$$

which is $o(E) + O(D)$ provided r goes to zero as E does.

We now turn to the terms in the two lines (89) and (90). Combining (43) with (99), we obtain by (93)

$$\int_{B_R} |\nabla\phi^r|^2 \lesssim \int_{\partial B_R} (\bar{g} - \bar{f})^2. \tag{102}$$

Hence in conjunction with (98) we learn that

$$\text{in line (90) we may replace } (\bar{g} - \bar{f})_r \text{ by } \bar{g} - \bar{f}, \tag{103}$$

once more at the expense of an error $r^{\frac{1}{2}} O(E + D)$. Now comes the beneficial effect of mollification: By standard regularity theory for the Neumann-Poisson problem, $\sup_{\bar{B}_R} |\nabla\phi^r|$ is estimated by a sufficiently high norm of $(\bar{g} - \bar{f})_r$, which due to the mollification is controlled by $(\int_{\partial B_R}(\bar{g} - \bar{f})^2)^{\frac{1}{2}}$. A closer inspection shows that in line with scaling, this estimate assumes the form

$$\sup_{\bar{B}_R} |\nabla\phi^r|^2 \lesssim \frac{1}{r^{d-1}} \int_{\partial B_R} (\bar{g} - \bar{f})^2 \overset{(93)}{=} \frac{1}{r^{d-1}} O(E + D). \tag{104}$$

Thus we may appeal to the following easy consequence of the definition of W^2 (and Cauchy-Schwarz)

$$\Big|\int_{\partial B_R} \phi^r (\bar{g} - dg)\Big| \le \sup_{\bar{B}_R} |\nabla\phi^r| \Big(\big(\int_{\partial B_R} \bar{g} + g(\partial B_R)\big) W^2(\bar{g}, g)\Big)^{\frac{1}{2}}, \tag{105}$$

and a corresponding estimate for f. Hence we learn from both (92) and (93) that the contribution from (103) is $r^{-\frac{d-1}{2}}(o(E) + O(D))$.

In conclusion, we obtain that the term in line (90) satisfies

$$\int_{\partial B_R} \phi^r ((\bar{g} - \bar{f})_r - d(g - f)) = \frac{1}{r^{\frac{d-1}{2}}} (o(E) + O(D)) + r^{\frac{1}{2}} O(E + D),$$

which still is $o(E) + O(D)$ provided r only slowly goes to zero as $E + D$ does. Appealing once more to an estimate similar to (105), but this time in conjunction with (96) and (97), we learn that the term in line (89) is

$$\int_{B_R} \phi^r \Big(\frac{\mu(B_R)}{|B_R|} - \frac{\lambda(B_R)}{|B_R|} - d(\mu - \lambda)\Big) = \frac{1}{r^{\frac{d-1}{2}}} O(D),$$

and thus is well-behaved, too.

It remains to address the term in line (91); we claim that

$$\int_\Omega \int_\sigma^\tau |\nabla\phi^r(X(t))|^2 dt d\pi - \int_{B_R} |\nabla\phi^r|^2$$
$$\leq \frac{1}{r^{d-1}} o(E+D) + \frac{1}{r^d}(E+D)^{\frac{3}{2}}. \tag{106}$$

Indeed, setting $\zeta = |\nabla\phi^r|^2$, and appealing to (1) and (6), we split the l. h. s. of (106) into the three differences

$$\int_0^1 \int_\Omega \big(I(X(t) \in \bar{B}_R)\zeta(X(t)) - I(X(0) \in B_R)\zeta(X(0))\big) d\pi dt$$
$$+ \Big(\int_{B_R} \zeta d\lambda - \kappa_R \int_{B_R} \zeta\Big) + (\kappa_R - 1)\int_{B_R} \zeta. \tag{107}$$

We turn to the first difference in (107) and note

$$I(X(t) \in \bar{B}_R)\zeta(X(t)) - I(X(0) \in B_R)\zeta(X(0))$$
$$\leq I\big(\exists s \in [0,1]\ X(s) \in \partial B_R, X(t) \in \bar{B}_R\big)\zeta(X(t))$$
$$+ I\big(\forall s \in [0,1]\ X(s) \in B_R\big)\big(\zeta(X(t)) - \zeta(X(0))\big),$$

so that it is $\leq$

$$\sup_{\bar{B}_R} \zeta\, \pi\big(\Omega \cap \{\exists s \in [0,1]\ X(s) \in \partial B_R\}\big) + \sup_{\bar{B}_R} |\nabla\zeta| \int_\Omega |x-y| d\pi. \tag{108}$$

According to (95), the second factor of the first term is $o(1)$. The second factor of the second term is $\leq (\pi(\Omega)\int_\Omega |x-y|^2 d\pi)^{\frac{1}{2}}$, and thus $O(E^{\frac{1}{2}})$ by definitions (12) and (13). Hence it remains to control the two first factors in (108). Recalling the definition of ζ and (104), we have $\sup_{\bar{B}_R} \zeta = r^{-(d-1)} O(E+D)$, so that the first term in (108) is $r^{-(d-1)} o(E+D)$. By the same argument that led to (104), we have

$$\sup_{\bar{B}_R} |\nabla^2\phi^r|^2 \lesssim \frac{1}{r^{d+1}} \int_{\partial B_R} (\bar{g} - \bar{f})^2 \overset{(93)}{=} \frac{1}{r^d} O(E+D), \tag{109}$$

which in combination with (104) yields

$$\sup_{\bar{B}_R} |\nabla\zeta| \lesssim \frac{1}{r^d} \int_{\partial B_R} (\bar{g} - \bar{f})^2 = \frac{1}{r^d} O(E+D), \tag{110}$$

so that the second term in (108) is $r^{-d} O(E^{\frac{1}{2}}(E+D))$.

We now address the second difference in (107). In analogy to (105), we have that it is $\leq$

$$\sup_{\bar{B}_R} |\nabla\zeta| \big(\lambda(B_R) + \kappa_R |B_R|\big) W^2(\lambda \llcorner B_R, \kappa_R dx \llcorner B_R)\big)^{\frac{1}{2}}.$$

Due to (70) in Lemma 4 and to (110) this is $r^{-d} O((E + D)D^{\frac{1}{2}})$. Finally, the third term in (107) is $\leq (\sup \zeta)|\kappa_R - 1|$ and thus $r^{-(d-1)} O((E + D)D^{\frac{1}{2}})$.

The final task left is to estimate the l. h. s. of (40) by the l. h. s. (87). We start by approximating the argument of $\nabla\phi^r$:

$$\int_\Omega \int_\sigma^\tau |\nabla\phi^r(X(t)) - I(x \in B_R)\nabla\phi^r(x)|^2 dt d\pi.$$

Using

$$\begin{aligned}
& I(t \in [\sigma, \tau]) \big|\nabla\phi^r(X(t)) - I(x \in B_R)\nabla\phi^r(x)\big)\big|^2 \\
& \overset{(6)}{=} \big|I(X(t) \in \bar{B}_R)\big(\nabla\phi^r(X(t)) - I(X(0) \in B_R)\nabla\phi^r(X(0))\big)\big|^2 \\
& \leq I\big(\exists s \in [0, 1]\ X(s) \in \partial B_R, X(t) \in \bar{B}_R\big)|\nabla\phi^r(X(t))|^2 \\
& + I\big(\forall s \in [0, 1]\ X(s) \in B_R\big)\big|\nabla\phi^r(X(t)) - \nabla\phi^r(X(0))\big|^2,
\end{aligned}$$

we obtain

$$\begin{aligned}
& \int_\Omega \int_\sigma^\tau |\nabla\phi^r(X(t)) - I(x \in B_R)\nabla\phi^r(x)|^2 dt d\pi \\
& \quad \leq \sup_{\bar{B}_R} |\nabla\phi^r|^2 \pi\big(\Omega \cap \{\exists s \in [0, 1]\ X(s) \in \partial B_R\}\big) \\
& \quad + \sup_{\bar{B}_R} |\nabla^2\phi^r|^2 \int_\Omega |x - y|^2 d\pi.
\end{aligned}$$

By definition (12) and (13), as well as (94), (104) and (109) this yields

$$\begin{aligned}
& \int_\Omega \int_\sigma^\tau |\nabla\phi^r(X(t)) - I(x \in B_R)\nabla\phi^r(x)|^2 dt d\pi \\
& \quad \leq \frac{1}{r^{d-1}} o(E + D) + \frac{1}{r^d} O((E + D)E).
\end{aligned} \tag{111}$$

By the triangle inequality we obtain from (111)

$$\begin{aligned}
& \int_\Omega (\tau - \sigma)|(y - x) - I(x \in B_R)\nabla\phi^r(x)|^2 d\pi \\
& \leq 2 \int_\Omega \int_\sigma^\tau |\dot{X}(t) - \nabla\phi^r(X(t))|^2 dt d\pi \\
& + \frac{1}{r^{d-1}} o(E + D) + \frac{1}{r^d} O((E + D)E).
\end{aligned}$$

By definition (12) of Ω and $R \geq 1$, we clearly have

$$\big((B_1 \times \mathbb{R}^d) \cup (\mathbb{R}^d \times B_1)\big) \cap \mathrm{supp}\pi \subset \Omega.$$

Because of $R \geq 2$ and in our regime of short trajectories, cf. (65), we have

$$\tau - \sigma = 1, \;\; x \in B_R \quad \text{for}\, (x, y) \in \big((B_1 \times \mathbb{R}^d) \cup (\mathbb{R}^d \times B_1)\big) \cap \mathrm{supp}\pi.$$

This implies

$$\begin{aligned}&\int_{(B_1 \times \mathbb{R}^d) \cup (\mathbb{R}^d \times B_1)} |(y - x) - \nabla\phi^r(x)|^2 d\pi \\ &\leq \int_\Omega (\tau - \sigma)|(y - x) - I(x \in B_R)\nabla\phi^r(x)|^2 d\pi.\end{aligned}$$

The combination of the two last inequalities connects the l. h. s. of (40) (with ϕ replaced by ϕ^r) with the l. h. s. (87). This concludes the proof of Proposition 1.

References

1. L. Ambrosio, F. Glaudo, Finer estimates on the 2-dimensional matching problem. J. Ec. Polytech. - Math. **6**, 737–765 (2019)
2. L.A. Caffarelli, A localization property of viscosity solutions to the Monge-Ampère equation and their strict convexity. Ann. Math. **131**(2), 129–134 (1990)
3. L.A. Caffarelli, The regularity of mappings with a convex potential. J. Am. Math. Soc. **5**(1), 99–104 (1992)
4. G. De Philippis, A. Figalli, Partial regularity for optimal transport maps. Publ. Math. Inst. Hautes Etudes Sci. **121**, 81–112 (2015)
5. A. Figalli, Y.-H. Kim, Partial regularity of Brenier solutions of the Monge-Ampère equation. Discret. Contin. Dyn. Syst. (Ser. A) **28**, 559–565 (2010)
6. M. Goldman, M. Huesmann, F. Otto, Quantitative linearization results for the Monge-Ampère equation. Comm. Pure Appl. Math. **74**, 2483–2560 (2021)
7. M. Goldman, F. Otto, A variational proof of partial regularity for optimal transportation maps. Ann. Sci. Éc. Norm. Supér. (4) **53**(5), 1209–1233 (2020)
8. M. Huesmann, F. Mattesini, F. Otto, There is no stationary cyclically monotone Poisson matching in 2d (2021). arXiv:2109.13590
9. F. Otto, M. Prod'homme, T. Ried, Variational approach to regularity of optimal transport maps: general cost functions. Ann. PDE 7 (2021)
10. R.A. Adams, J.J. F. Fournier, *Sobolev Spaces* (Academic, 2003)
11. F. Santambrogio, *Optimal Transport for Applied Mathematicians* (Birkhäuser, 2015)

The Optimal Transport Principle for Developmental Biology

Geoffrey Schiebinger

Abstract Biology has entered a new era of precision measurement and massive datasets. Techniques like single-cell RNA sequencing (scRNA-seq) and single-cell ATAC-seq have emerged as powerful tools to profile cell states at unprecedented molecular resolution. One of the most exciting prospects associated with this new trove of data is the possibility of studying temporal processes, such as differentiation and development. If we could understand the genetic forces that control embryonic development, then we would have a better idea of how cell types are stabilized throughout adult life and how they destabilize with age or in diseases like cancer.

This would be within reach if we could analyze the dynamic changes in gene expression, as populations develop and subpopulations differentiate. However, this is not directly possible with current measurement technologies because they are destructive (e.g. cells must be lysed to measure expression profiles). Therefore, we cannot directly observe the waves of transcriptional patterns that dictate changes in cell type. In response, there has been a flurry of recent work on developing methods to infer trajectories from static snapshots of gene expression profiles (e.g. [1], [2], [3], [4]). However, there is relatively little theoretical understanding of this statistical inverse problem; if we are to rely on trajectory inference to understand disease, develop new therapies, and engineer tissues, we need to know when to trust the results.

This short course introduces a rigorous framework for understanding the developmental trajectories of cells in a dynamically changing, heterogeneous population based on static snapshots along a time-course. The framework is based on a simple hypothesis: over short time-scales cells can only change their expression profile by small amounts. We formulate this in precise mathematical terms using a classical tool called optimal transport (OT), and we propose that this optimal transport hypothesis is one of the first fundamental mathematical principles of developmental biology. Compared to related fields like evolution and population genetics, developmental biology has been relatively non-mathematical. This OT-hypothesis leads

G. Schiebinger (✉)
University of British Columbia, Vancouver, BC, Canada
e-mail: geoff@math.ubc.ca

© The Author(s), under exclusive license to Springer Nature Switzerland AG 2026
Y. Kim et al. (eds.), *The Mathematics of Monge-Kantorovich Optimal Transport*, Springer Proceedings in Mathematics & Statistics 552,
https://doi.org/10.1007/978-3-032-23146-8_3

to a rigorous mathematical theory of development, broadly interpreted to include any population of cells changing over time (e.g. tumorigenesis, disease progression, aging, wound healing, cellular reprogramming etc).

1 Introduction

The human body is made of trillions of cells. From skin cells, to muscle cells, to neurons, the cell is the fundamental unit of organization of every organ in the body. Within each cell, DNA stores the genetic code in a meter-long string of 3 billion nucleotide letters (ACTG) called the *genome*. There are roughly 20,000 distinct parts of the genome called genes, which can be transcribed into RNA. While the DNA is protected inside the cellular nucleus, the RNA transcripts exit the nucleus, travel around the cell, and perform various functions like being translated into proteins.

In a given cell, only a small fraction of these 20,000 genes will be *expressed*, or actively used to create RNA transcripts and proteins. Therefore, while each cell has the same DNA, different cell types express different genes in order to achieve different functions. Skin cells express certain genes to make collagen and melanin; neurons express other genes to make neurotransmitters.

1.1 Single Cell RNA Sequencing Reveals the Vectors of Cell Identity

The RNA transcripts can be captured from cells, converted back into DNA, and loaded into a DNA sequencing machine which reads out the sequence of nucleotide letters for each transcript. The sequence of each transcript can then be mapped to a specific location of the genome to reveal which genes the cells expressed. More precisely, for each of the 20,000 genes, one can count how many transcripts were detected to form a 20,000 dimensional **gene expression vector**. The value of the ith entry of this vector encodes the number of molecules of RNA detected for the ith gene.

Modern single-cell RNA-sequencing (scRNA-seq) technologies employ can measure gene expression vectors for millions of cells. This is achieved through a clever strategy called DNA barcoding. While it is not possible to perform separate DNA sequencing reactions on each cell, the RNA transcripts from individual cells are labeled with distinct *DNA barcodes*, illustrated in Fig. 1. Then, transcripts from many different cells can be pooled together, sequenced together, and then computationally separated to form separate gene expression vectors. Without these DNA barcodes, one would simply measure a single vector corresponding to the sum of all gene expression vectors from all the cells.

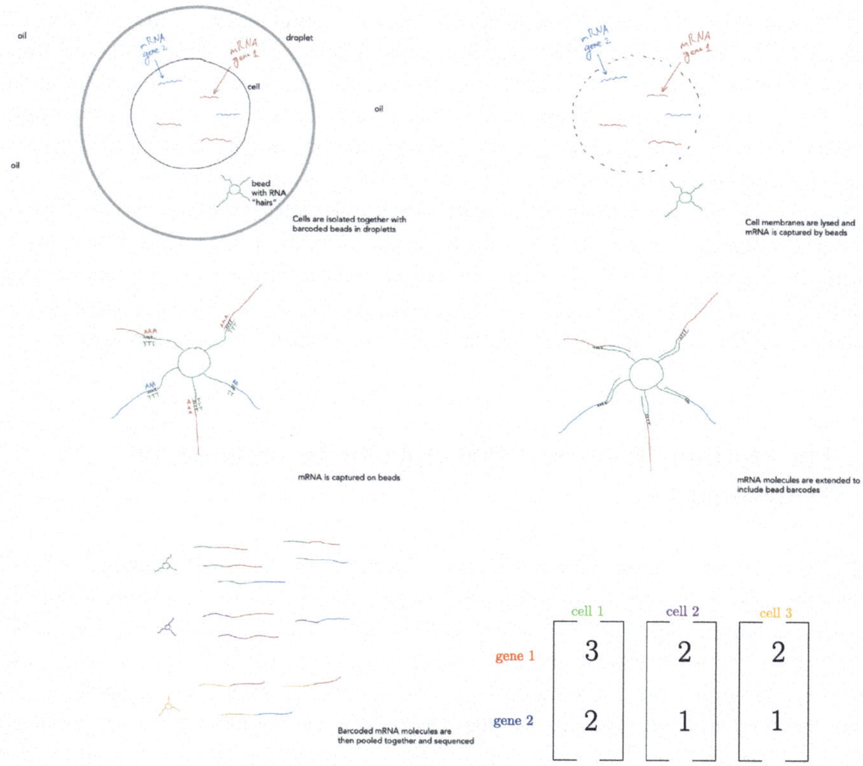

Fig. 1 DNA barcodes enable single cell RNA sequencing. (a) Cells are isolated in dropletts of fluid surrounded by oil. Inside the droplet, there is also a bead coated with many DNA strands. (b) Cells are lysed, meaning the membrane is dissolved, (c) the RNA from the cell is captured on the bead, (d) the RNA is extended to incorporate a DNA "barcode" unique to each cell. (e) RNA from different cells can be pooled together for sequencing. (f) When the sequence of each RNA molecule is read, we keep track of how many copies of RNA we see from each cell for each gene and store these in a gene expression matrix

There are diverse strategies for labeling cells with cell-specific DNA barcodes. Early strategies involved isolating individual cells in miniature wells. In each well, a separate chemical reaction would lyse (or open) cells, capture the RNA, and then attach DNA barcodes. In a hard day's work, one could process about 100 cells with this type of approach. In 2015, several groups developed microfluidics-based technologies which could easily isolate tens of thousands of cells in droplets of fluid surrounded by oil [9]. Inside the droplet, the same chemical reactions can be performed to capture RNA and attach DNA barcodes, illustrated in Fig. 1. More recently, split-pool barcoding strategies have made it possible to profile millions of cells in a single experiment [11].

This DNA barcoding strategy has also fueled the development of numerous other single-cell measurement technologies, each of which can measure a different attribute of cell state. For example, ATAC-seq can measure which parts of the genome are accessible by attacking the DNA with transposase, an enzyme which cuts out fragments of DNA. Sequencing these fragments reveals which parts of the genome are *open* (or accessible) and which parts are *closed* (or inaccessible). This can shed light on how different cell types package the DNA inside the nucleus, closing off portions of the genome they don't need.

In this course, we will think of cell states abstractly as elements of a high-dimensional vector space, $\mathcal{X}$. We will represent a population of cells with a probability distribution P on $\mathcal{X}$. Finally, single-cell measurement technologies provide samples from P. For concreteness, we will often refer to $\mathcal{X}$ as *gene expression space*, because scRNA-seq is the most popular single-cell measurement technology.

2 The Optimal Transport Principle for Developmental Trajectories

Cell states are not static. Genes can turn other genes on or off, causing cells to move through gene expression space. For nearly a century, the prevailing theory of developmental biology has been based on Waddington's idea of an 'epigenetic landscape' [18]. In this classical metaphor, cells traverse a potential-energy landscape, traveling through valleys separated by energy barriers which prevent spontaneous conversion of one cell type into another (Fig. 2). If we could understand the shape of these landscapes and the underlying regulatory programs, then we would better

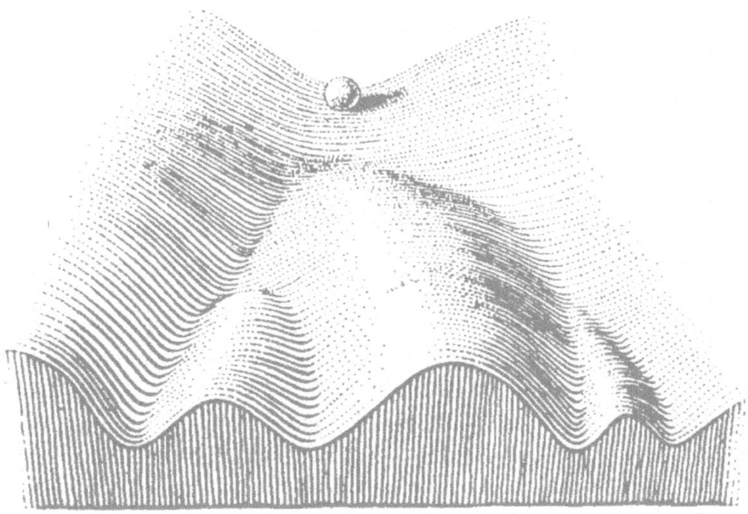

Fig. 2 Waddington's landscape

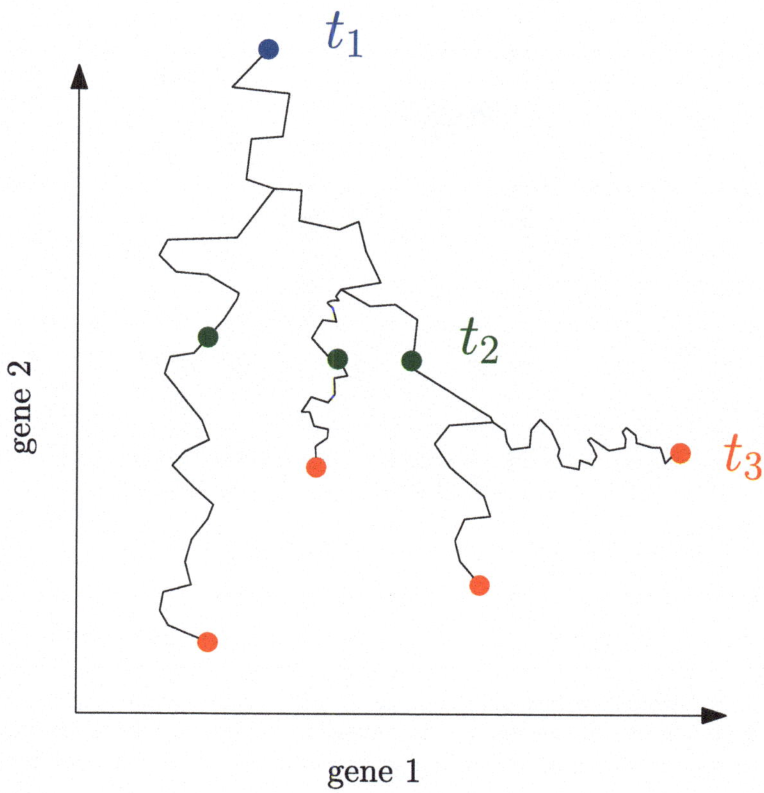

Fig. 3 A single blue cell at time t_1 gives rise to four descendants at time t_3. The black curves denote the path traversed as cells change gene expression over time. Cell division creates branching paths

understand how cell types are stabilized in normal development, how they destabilize with age or in diseases like cancer [5, 17, 19], and how cells can be reprogrammed [1].

This might be within reach if we could analyze the dynamic changes in gene expression, as populations develop and subpopulations differentiate. However, this is not directly possible with current measurement technologies because they are destructive (e.g. RNA must be extracted from cells to be sequenced). Therefore, we cannot directly observe the waves of transcriptional patterns that occur along cellular development, i.e. *developmental trajectories*.

As cells change state over time, they trace out paths in $\mathcal{X}$. Cellular division creates branching paths (Fig. 3). These paths are called developmental trajectories. Because scRNA-seq kills cells, we cannot directly observe developmental trajectories; we can only recover samples of static cell states. We can, however leverage the reproducibility of development to sample along various time-points of a single developmental process. For example, we can fertilize several mouse embryos at time 0, and collect cells from the first at time t_1, the second at time t_2 and so on and measure cell states with scRNA-seq (Fig. 4).

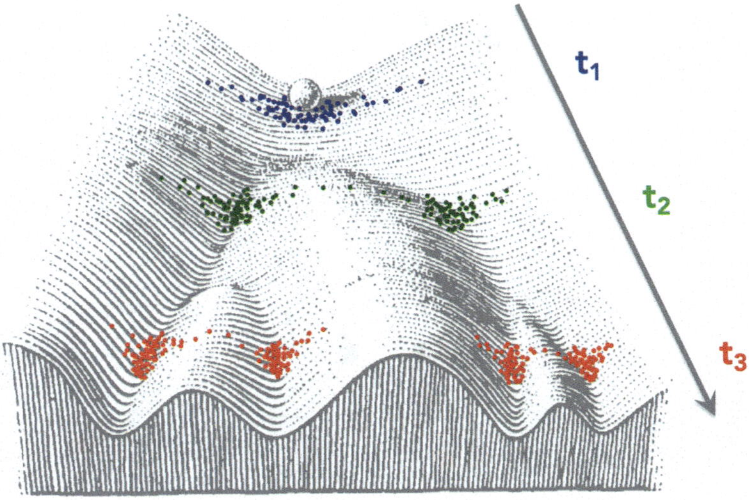

Fig. 4 Cells sampled at various time-points along a developmental time-course

We model the population of cells at time t with a probability distribution $\mathbb{P}_t$ (Fig. 5). As $\mathbb{P}_t$ changes over time, it describes a *developmental stochastic process*. Formally, a developmental stochastic process is a law on developmental trajectories. Any population of cells changing over time can be modeled with a developmental process. Examples of developmental processes include embryonic development, tumoregenesis, and wound healing.

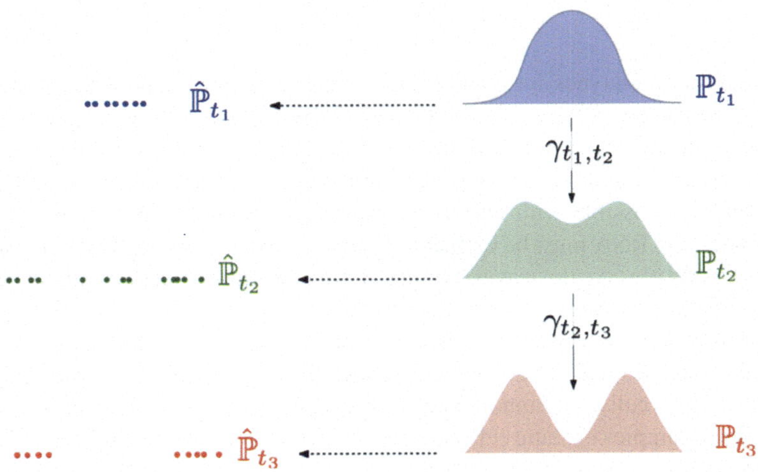

Fig. 5 A population of cells is modeled with a time-varying distribution $\mathbb{P}_t$ at time t. A Markov transition kernel γ links pairs of time-points. Single cell RNA-sequencing produces empirical distributions $\hat{\mathbb{P}}_{t_i}$

We generally assume that developmental processes are Markov. There are, then, Markov transition kernels $\gamma_{t_i,t_{i+1}}$ connecting the pairs of measured time-points. These transition kernels describe cellular dynamics: e.g. given a cell has state X_{t_1} at time t_1, what states could it obtain at time t_2? Sampling with scRNA-seq at time t_i yields an empirical distribution $\hat{\mathbb{P}}_{t_i}$. However, the samples from different time-points are independent, because they are collected from different embryos.

The goal of *trajectory inference* is to recover the law of a developmental process from such samples. Without further assumptions, this is of course impossible. By accumulating many samples, we can recover the fixed-time marginals, but how can we connect these together to recover trajectories?

One natural assumption is that individual cells do not move large distances in gene expression space over short time-scales. For example, suppose we observe a population of cells at time 0 with distribution $\mathbb{P}_0 = N(0, \sigma^2)$ and at time 1 we observe a shifted population with distribution $\mathbb{P}_1 = N(\mu, \sigma^2)$. Let $X_0 \sim \mathbb{P}_0$ and let $X_1 \sim \mathbb{P}_1$ denote cell states at time 0 and 1. How can we recover the joint distribution of X_0 and X_1? There are many possible joint distributions of X_0 and X_1, but if X_0 is *close* to X_1, then one natural idea is to look for the joint distribution minimizing the distance between X_0 and X_1

$$\hat{\gamma} \quad \leftarrow \quad \arg\min_{\pi} \mathbb{E}_{\pi} \|X_0 - X_1\|^2. \tag{1}$$

This has deterministic solution $X_1 = X_0 + \mu$.

This idea is related to a classical mathematical tool called *optimal transport* (OT). OT was first proposed by Gaspard Monge in the 1780s to efficiently relocate piles of earth (Monge, 1781; Kantorovich 1941; Villani, 2008). Today, OT is used in diverse areas of science and engineering, from fluid mechanics (Benamou and Brenier, 2000) to climate science (Ambrosio et al., 2012) to charting the distribution of mass in the early universe (Frisch et al., 2002) to pure math (Figalli, 2018).

2.1 The Optimal Transport Principle

A *transport plan* π specifies the amount of dirt transported from source location x to target location y: the amount we move is $\pi(x, y)$. If it costs $c(x, y)$ to send a unit mass from x to y, and we move an amount $\pi(x, y)$, then the total cost is given by

$$\mathbb{E}_{\pi} c(x, y) = \iint c(x, y)\pi(x, y)dxdy.$$

The optimal transport plan for redistributing a source distribution $\mathbb{P}$ to a target distribution $\mathbb{Q}$ is then given by minimizing the transport cost subject to constraints which enforce conservation of mass

$$\begin{aligned} \underset{\pi}{\text{minimize}} \quad & \iint c(x, y)\pi(x, y)dxdy \\ \text{subject to} \quad & \int \pi(x, y)dy = \mathbb{P}(x) \\ & \int \pi(x, y)dx = \mathbb{Q}(y) \end{aligned} \tag{2}$$

Here the first constraint says that if we add up the mass transported to all target destinations y, we obtain the source distribution $\mathbb{P}$. Conversely, the second constraint says that if we add up the mass transported from all source positions x, we obtain the target distribution $\mathbb{Q}$. This optimization problem is a linear program: the objective and constraints are both linear in the optimization variable π.

Here, we propose the *optimal transport principle* for developmental stochastic processes [7, 15]: over short time-scales the Markov transition kernels agree with the optimal transport coupling of the fixed-time marginals (Fig. 6). The OT hypothesis can be understood intuitively and geometrically: the curve traced out by the marginals $\{\mathbb{P}_t\}_t$ should be *locally linear* (Fig. 7). We call this a *developmental curve*. When we sample cells with scRNA-seq, we obtain empirical distributions $\hat{\mathbb{P}}_{t_i}$ which form noisy points along the developmental curve. The number of cells sampled determines the noise-level of these empirical distributions: as we sample more cells they lie closer to the curve.

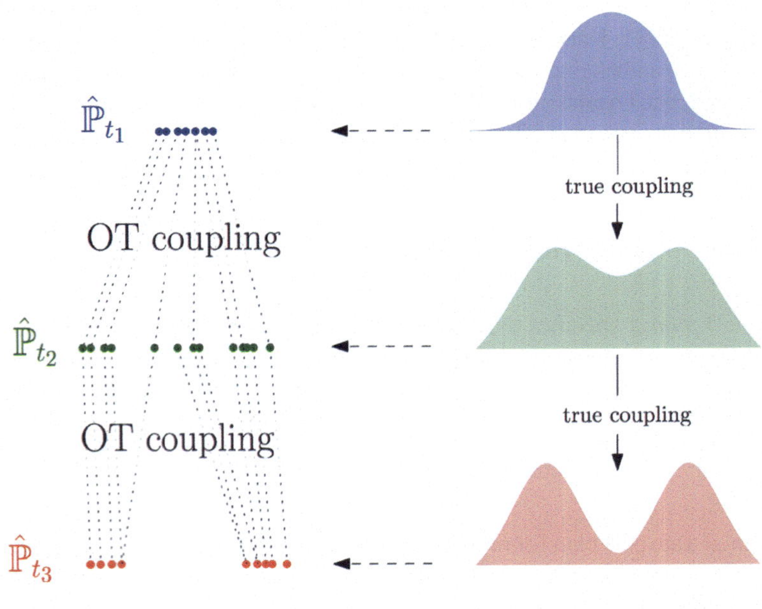

Fig. 6 According to the OT principle, the true Markov transition kernel agrees with optimal transport over short time-scales

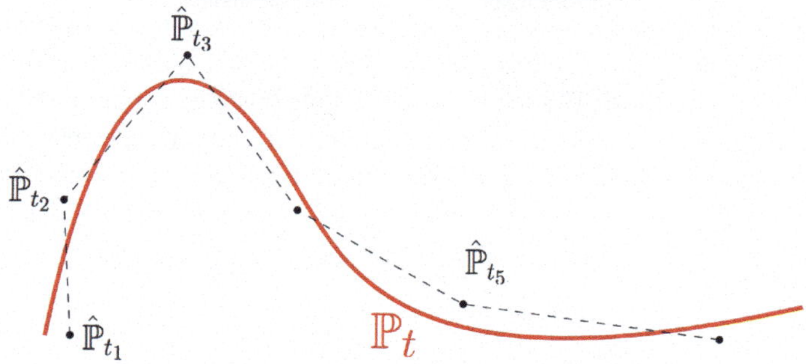

Fig. 7 A population of cells developing over time traces out a curve in the space of probability distributions (red curve). When we sample with single cell RNA-seq, we obtain empirical distributions at various time-points (black dots). We can then connect the empirical distributions with optimal transport to recover a piece-wise geodesic estimate of the curve (dashed black lines)

However, in order to state the OT hypothesis more precisely, we need to handle two biological considerations. First, classical OT conserves mass, but cells can proliferate. Second, as a linear program, OT typically gives a deterministic coupling. But cells can be partially fated.

Incorporating growth In order to handle cell growth, we introduce a growth function $g : \mathcal{X} \to \mathbb{R}$ that scales transported mass. This growth function can be defined in terms of the developmental stochastic process as follows. Let $Y_{t_2} \sim \mathbb{P}_{t_2}$ denote a random cell at time t_2. Let A_{t_1} denote its ancestor at time t_1. Then let γ_{t_1,t_2} be the joint distribution of (A_{t_1}, Y_{t_2}). Note that the t_1 marginal of γ_{t_1,t_2} can be different from $\mathbb{P}_{t_1}$. This difference is due to cell growth (and death). In particular, the marginals of γ_{t_1,t_2} are

$$\int \gamma_{t_1,t_2}(x, y)dx = \mathbb{P}_{t_2}(y)$$

$$\int \gamma_{t_1,t_2}(x, y)dy = \mathbb{P}_{t_1}(x)g(x)^{t_2-t_1}$$

for some growth function g with $\int g(x)^{\Delta_t} d\mathbb{P}_{t_1}(x) = 1$.

In practice, we can estimate this growth function from gene expression levels of cells [15]. We can then set up an optimal transport problem between a rescaled $\hat{\mathbb{P}}_{t_1}$ and $\hat{\mathbb{P}}_{t_2}$.

$$\begin{aligned} \hat{\pi}_{t_1,t_2} \quad = \quad \underset{\pi}{\operatorname{argmin}} \quad & \iint c(x, y)\pi(x, y)dxdy \\ \text{subject to} \quad & \int \pi(x, y)dy = d\hat{\mathbb{P}}_{t_2}(y) \\ & \int \pi(x, y)dx = d\hat{\mathbb{P}}_{t_1}(x)\hat{g}(x)^{\Delta_t} \end{aligned} \tag{3}$$

This rescaling can be understood as giving each cell alive at t_1 a certain number of tokens, proportional to the number of living descendants it is expected to have at t_2. Then these tokens are redistributed to obtain the observed distribution $\hat{\mathbb{P}}_{t_2}$.

Finally, because the growth rates are not known perfectly, we can use *unbalanced transport* to allow some flexibility in enforcing the marginal constraints involving growth:

$$\begin{aligned} \hat{\pi}_{t_1,t_2} \quad = \quad \underset{\pi}{\text{argmin}} \quad & \iint c(x,y)\pi(x,y)dxdy - \epsilon\mathcal{H}(\pi) \\ \text{subject to} \quad & \int \pi(x,y)dx \approx d\hat{\mathbb{P}}_{t_2}(y) \\ & \int \pi(x,y)dy \approx d\hat{\mathbb{P}}_{t_1}(x)\hat{g}(x)^{\Delta_t} \end{aligned}$$

For a precise formulation of unbalanced transport, see [3], and for its use in trajectory inference see [15].

Entropy regularized OT The second modification we make is to add entropic regularization. This has several practical benefits which we describe here, in addition to theoretical benefits which we describe later in Sect. 4. The first benefit of entropic regularization is algorithmic: it makes the objective function strongly convex and the optimization problem can then be solved much faster compared to the linear program version.

Second, it allows to model partially fated cells. As a linear program, ordinary optimal transport produces a solution at an "extreme point", which in the case of couplings corresponds to a matching (or permutation). Without entropy regularization, each cell would be connected to a single descendant at later time-points and a single ancestor at earlier time-points, as shown in Fig. 8a.

Entropy-regularized OT trades off minimizing transport cost with maximizing the entropy of the transport plan $\mathcal{H}(\pi) = \int \pi \log \pi$. The level of trade-off is controlled by a regularization parameter ϵ as follows

$$\begin{aligned} \hat{\pi}_{t_1,t_2} \quad = \quad \underset{\pi}{\text{argmin}} \quad & \iint c(x,y)\pi(x,y)dxdy - \epsilon\mathcal{H}(\pi) \\ \text{subject to} \quad & \int \pi(x,y)dy = d\hat{\mathbb{P}}_{t_2}(y) \\ & \int \pi(x,y)dx = d\hat{\mathbb{P}}_{t_1}(x)\hat{g}(x)^{\Delta_t} \end{aligned} \tag{4}$$

This has the effect of pushing the optimal π away from perfect matchings, which have zero entropy. Each cell is then connected to multiple potential descendants at later time-points and ancestors at earlier time-points, as visualized in Fig. 8a. The resulting transport plan $\hat{\pi}$ can be used to estimate the ancestors and descendants of cells across time-points. There is a row of $\hat{\pi}_{t_1,t_2}$ for each cell x_i from time t_1 and a column for each cell y_j from time t_2. The i, j entry of the matrix $\hat{\pi}_{t_1,t_2}$ encodes the amount of mass transported from x_i to y_j. The ith row of $\hat{\pi}_{t_1,t_2}$ specifies the

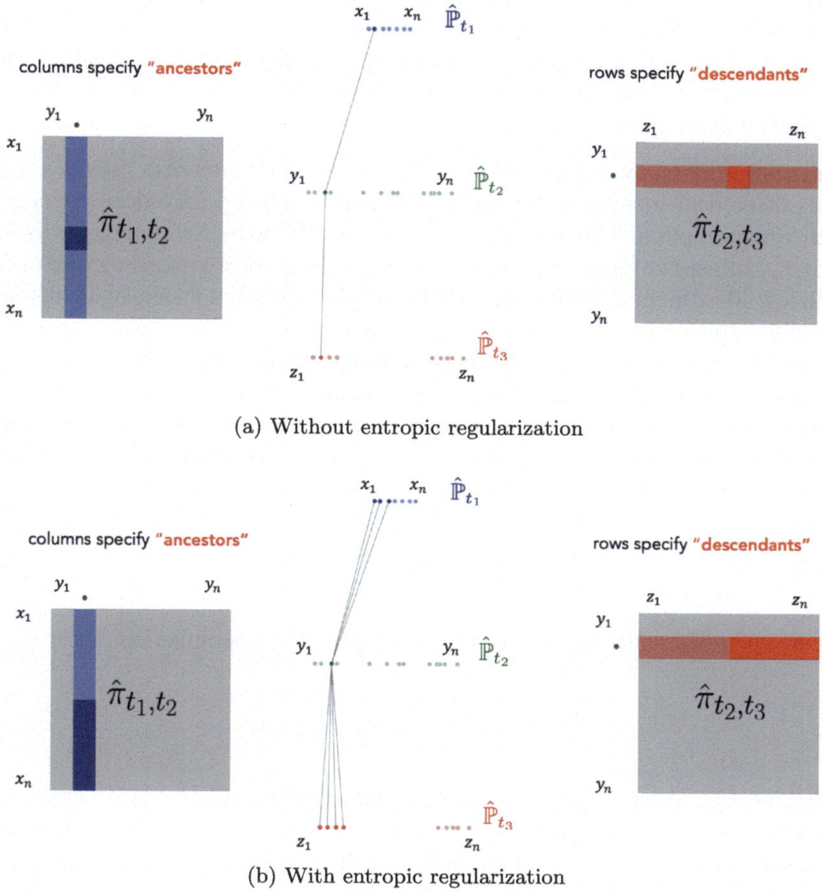

Fig. 8 The effect of entropy regularization on ancestor and descendant distributions. Cells sampled at 3 time-points are shown with dots (center). Cells from t_1 are denoted $x_1, \ldots, x_n$, cells from time t_2 are denoted $y_1, \ldots, y_n$ and cells from time t_3 are denoted $z_1, \ldots, z_n$. The coupling matrix connecting time t_1 to time t_2 is shown on the left and the coupling matrix connecting time t_2 to time t_3 is shown on the right. A specific cell y_i is highlighted from time t_2 and its ancestor and descendant distributions are shown with and without entropy. The ancestor distribution of y_i (blue column of $\hat{\pi}_{t_1,t_2}$ on the left) is distributed over few cells in (a) without entropic regularization and over many cells in (b) with entropy regularization. Similarly, the descendant distribution (red row of $\hat{\pi}_{t_2,t_3}$ on the right is distributed over few cells in (a) without entropy regularziation and over many cells in (b) with entropy regularization

descendant distribution of cell x_i. This descendant distribution is a probability distribution over cells y_j specifying the probability that x_i transitions to y_j. Similarly, the jth column of $\hat{\pi}_{t_1,t_2}$ specifies the ancestor distribution of cell y_j. A larger value for the entropy regularization parameter ϵ encourages more entropic ancestor and descendant distributions.

Interpreting entropy regularization Entropic regularization also provides a rigorous theoretical foundation for trajectory inference, based on a deep connection to Brownian motion. This was discovered in the 1930s by Schrodinger who was studying diffusion of indistinguishable particles. Suppose we take a picture of two particles, allow them to diffuse for a short period of time and then take another picture of their positions Fig. 9. If the particles are indistinguishable, we don't precisely know which particle in the first picture corresponds to each particle in the second picture. There is a probability distribution over matchings, which we can easily compute using the fact that Brownian motion has Normal increments. Let $a, b \in \mathbb{R}^2$ denote the particle positions at time 0 and let x, y denote the observed positions at time 1. Let M_1 be the matching $a \to x$ and $b \to y$. Then the probability of M_1 is

$$p_1 = \frac{1}{\sqrt{2\pi}} e^{-\frac{\|a-x\|^2}{2\sigma^2} - \frac{\|b-y\|^2}{2\sigma^2}}.$$

The two possible matchings can be represented by 2×2 permutation matrices

$$M_1 = \begin{bmatrix} 1 & 0 \\ 0 & 1 \end{bmatrix} \quad M_2 = \begin{bmatrix} 0 & 1 \\ 1 & 0 \end{bmatrix} \tag{5}$$

Schrodinger showed that if we compute entropically regularized OT between the measures $\mu = \frac{1}{2}\delta_a + \frac{1}{2}\delta_b$ and $\nu = \frac{1}{2}\delta_x + \frac{1}{2}\delta_y$, and for the entropic regularization parameter we use ϵ, the diffusion coefficient of the Brownian motion, then the output is equal to the expected matching, $E = p_1 M_1 + p_2 M_2$.

Theorem 1 *Let* $\pi^\star = \arg\min_\pi \sum_{i \in \{a,b\}} \sum_{j \in \{x,y\}} \pi_{i,j} \|i - j\|^2 + \epsilon \mathcal{H}(\pi)$. *Then* $\pi^\star = p_1 M_1 + p_2 M_2$.

For a proof of this theorem, see Christian Leonard's lecture notes [8].

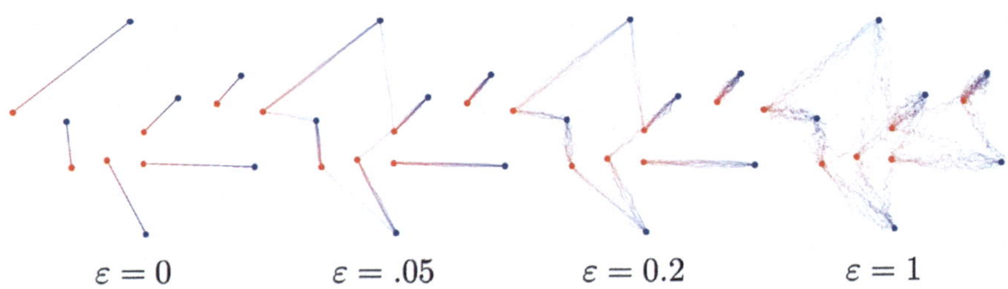

Fig. 9 Paths traveled by diffusing particles at various entropy levels

In Sect. 4 we leverage this result to establish a connection between the OT principle and Waddington's landscape. In essence, the OT principle is equivalent to assuming that developmental trajectories are paths of steepest descent down some Waddington landscape. However, due to thermodynamic fluctuations (e.g. in the diffusion of RNA within the cell), we also add some diffusion and model trajectories with a stochastic differential equation (SDE)

$$dX_t = -\nabla\Phi(X_t) + \epsilon dB_t, \tag{6}$$

where X_t denotes a cell state at time $t \in (0, 1)$, Φ is Waddington's landscape, and B_t is Brownian motion with diffusion coefficient ϵ. On short time-scales, the diffusion due to Brownian motion dominates drift down Waddington's landscape because diffusion is order $\sqrt{dt}$ and drift is order dt Therefore, on short time-scales the trajectories agree with entropic OT.

Given a single-cell RNA-seq time-course dataset, we are presented with the following statistical inverse problem: can we recover the law on paths of the SDE (6) by sampling the fixed-time marginals at various time-points? A direct approach, where we somehow parameterize the landscape Φ and search for the best parameters to fit data, would inevitably lead to a non-convex optimization problem. We will see in Sect. 4 that the OT principle leads to a consistent estimator readily computable through convex optimization. Ultimately, a deep understanding of this statistical inverse problem would shed light on the optimal design of time-course experiments. If the biological community is to rely on trajectory inference to understand disease, develop new therapies, and engineer tissues, we need to know when to trust the results.

3 Example Applications: Cellular Reprogramming and Urchin Embryonic Development

In this section, we explore several examples of applying the optimal transport principle to infer developmental trajectories from static snapshots. We have recently tested the OT hypothesis in diverse systems including induced pluripotent stem cell (iPSC) reprogramming in mice [1], sea urchin embryonic development [6], Arabidopsis root growth [7], [8], [9], human hematopoiesis [10], and reprogramming human cells. Other groups have recently applied OT to study high-plasticity states in lung cancer evolution [11], lineage plasticity in distal lung progenitors [12], and trajectories of aging [13].

Cellular reprogramming In this section, we illustrate an example application of OT to infer trajectories in a time-course of cellular reprogramming [15]. This is the dataset I analyzed during my postdoc at the Broad Institute, where I first proposed the optimal transport principle. We were studying the process of reprogramming mouse cells from fibroblasts into *induced pluripotent stem cells* (iPSCs). This process works by turning on genes to push the fibroblasts through gene expression space until they obtain a stem-cell state. This process was discovered in 2007 by Yamanaka, who was awarded the Nobel prize in medicine [16]. While reprogramming may ultimately hold potential for personalized medicine, the process by which cells reprogram is still not perfectly understood.

We prepared 38 identical batches of cells on day 0. Then, to start the reprogramming process, a chemical was added to turn on various genes and push cells through gene expression space. We then sampled cells with scRNA-seq at 12 hour resolution for a total of 38 time-points (there was a short window where we collected additional data to obtain finer resolution). Overall, we collected 250,000 scRNA-seq profiles. These data are visualized in Fig. 10.

While the starting population of fibroblasts is relatively pure, eventually a mixture of different cell types emerges including stem cells, as desired, but also other cell types such as neural and placental cells. We used optimal transport to connect cells sampled across time-points. The resulting trajectories (i.e. the sequence of ancestor distributions) are visualized in Fig. 11.

We validated these trajectories in several different ways. First, we tested whether we could use *geodesic interpolation* to interpolate the distribution of cells at held out time-points (Fig. 12). We found that geodesic interpolation could predict held-out intermediate time-points significantly better than by chance [15]. (In the early time-points, when there is essentially only one cell type, there is really only one possible coupling.)

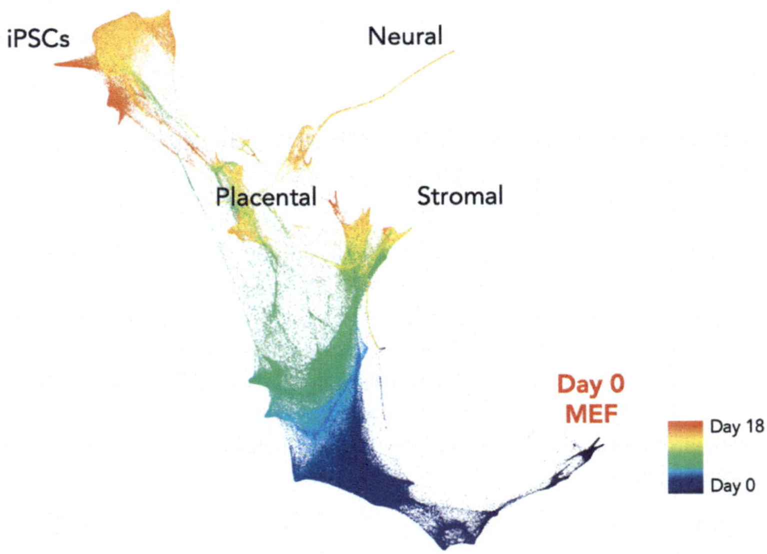

Fig. 10 Each dot shows a 20,000 dimensional gene expression vector, visualized in 2 dimensions through a technique called force-layout embedding. Cells are colored by time of collection

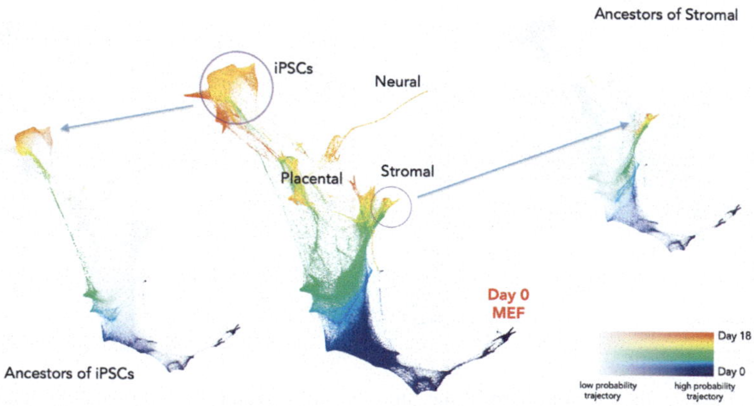

Fig. 11 Trajectories of reprogramming are visualized in two dimensions. As in Fig. 10, each dot is a single cell and cells are colored by time of collection. But now cell opacity encodes the probability of being an ancestor of the selected cell set. Bright colors indicate high-probability values of the ancestor distribution, fading to grey indicates low probability values of the ancestor distribution

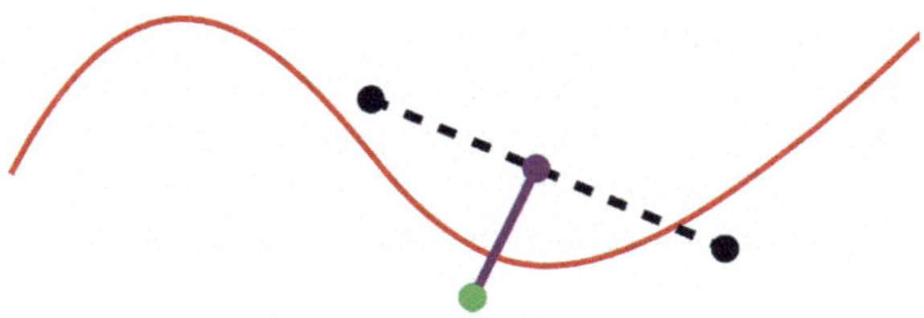

Fig. 12 One can test the performance of OT by holding out a time-point (green) and comparing to the mid-point (purple) of a geodesic interpolation

Next, we analyzed patterns of gene expression along different trajectories in order to identify genes pushing cells towards the stem-cell like state. We identified several candidate genes, tested them experimentally, and found that turning them on at day 0 led to a higher percentage of cells ending up as stem cells (rather than neural, placental, etc) [15].

Finally, we also looked for potential cell-cell signaling between stromal cells, which are naturally supportive, and the stem cells. In order to identify formative interactions, we looked at the ancestors of stromal cells and iPSCs at earlier time-points. We looked for simultaneous expression of ligands (signaling molecules) in stromal ancestors and the corresponding receptors in stem cell ancestors. We experimentally verified that the top predicted interaction was in fact real – adding more ligand increased the efficiency of reprogramming [15].

If you would like to explore this dataset yourself, the data and python notebooks are available for download at the WaddingtonOT Tutorial website [14].

Sea urchin embryonic development We have also seen encouraging results from the OT principle in sea urchin embryonic development. In collaboration with Greg

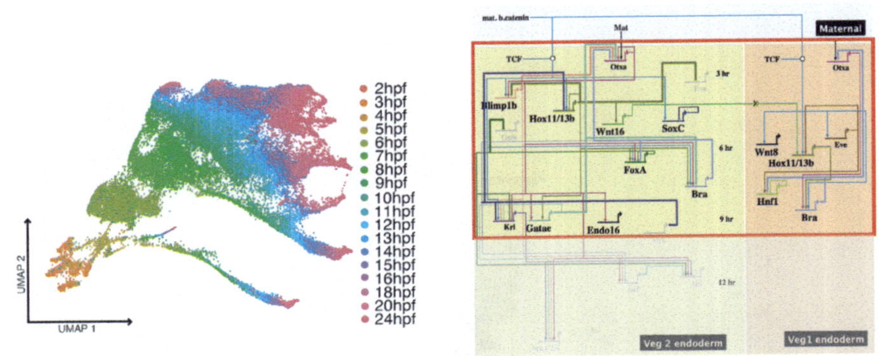

Fig. 13 A time-course of urchin embryonic development (left) and a reconstructed gene regulatory network (right). Similar to an electrical circuit, genes are connected by network of wires showing which genes turn others on or off. Gene names written in black are correctly recovered by optimal transport analysis

Wray and Dave McClay from Duke University, we collected 20 time-points over the first 24 hours of urchin embryonic development. We visualize the high-dimensional gene expression data in Fig. 13. Using the OT principle, we were able to rediscover most of the key genes regulating urchin development. For example, Fig. 13 shows the gene regulatory network for endoderm where OT trajectories correctly identified 18 out of 21 genes.

3.1 Developmental Curves and the Optimal Design of Experiments

Through testing the OT principle in diverse biological settings, we have gained some perspective on the optimal design of time-course experiments for analyzing developmental trajectories. In general, there is a trade-off between sampling at high temporal resolution and fully profiling gene expression at any specific instance. When we sample cells at a time-point with scRNA-seq, the empirical distribution of cells forms a "noisy measurement" of a point along the developmental curve (Fig. 14). The number of cells sampled determines the "noise-level" of the time-point measurement: the more cells we sequence at a particular developmental time, the more precisely we can localize the curve at that one position along the developmental curve. For a fixed budget of cells, profiling more time-points comes at the cost of fewer cells per time-point.

A key outcome of this viewpoint is a paradigm shift in the design of experiments: while the number of cells per study has increased dramatically with droplet-based scRNA-seq [14], [15], and split-pool barcoding [12], the number of time points in time-course studies of development has not increased by nearly the same amount.

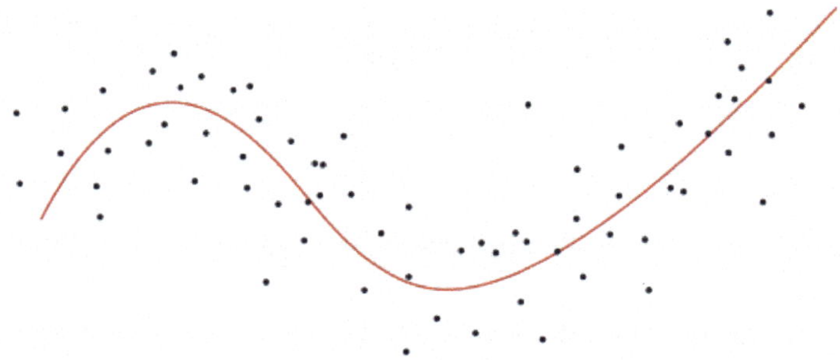

Fig. 14 As in Fig. 7, a population of cells developing over time traces out a curve in the space of probability distributions (red curve). When we sample with single cell RNA-seq, we obtain empirical distributions at various time-points (black dots). Here we illustrate a time-course dataset with more time-points and fewer cells per time-point, compared to Fig. 7

For example, a recent high-profile study on mouse embryonic development profiled an impressive one million cells, but over only 5 developmental time-points [16]. While high-resolution sampling of many time-points can be practically very difficult, the theory we present in Sect. 4 recommends collecting more time-points with fewer cells per time- point. To illustrate this, we analyzed the data from iPSC reprogramming [15] and sea urchin embryonic development [10]. Similar to analyzing saturation of reads in sequencing, we examined the saturation levels for both temporal resolution and numbers of cells (Fig. 15). We found that geodesic interpolation

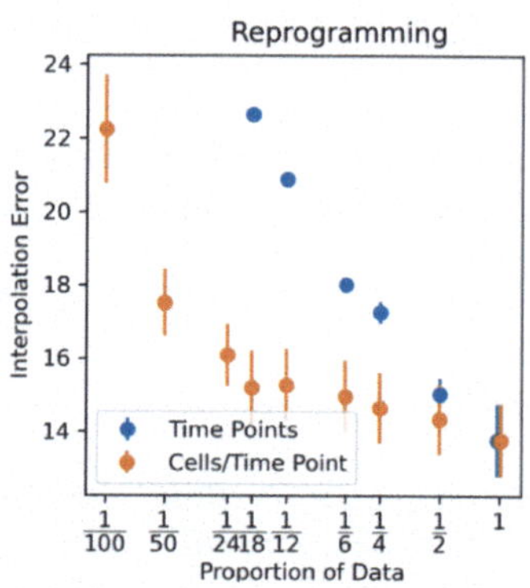

(a) Saturation of reprogramming time-course

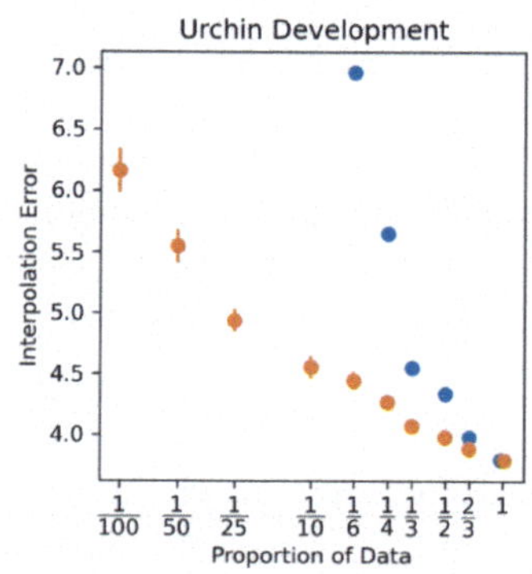

(b) Saturation of urchin time-course

Fig. 15 (a) (missing) Development traces out a curve in the space of probability distributions. Black dots indicate sampled populations at various time- points, which we connect using optimal transport (dashed lines). (b) Geodesic interpolation compares held out data (green) to mid-point of segment (purple) to quantify performance. (c) Saturation analysis for iPSC reprogramming and urchin embryonic development. Subsampling time-points (blue) causes interpolation to degrade more quickly compared to subsampling cells (orange)

results degraded more rapidly when time-points were removed, compared to down-sampling cells [13]. Therefore, it seems better to collect more time-points, with fewer cells per time-point.

4 A Convex Variational Approach to Trajectory Inference

In this section, we present some theoretical results on the statistical inverse problem of trajectory inference. We model cell developmental trajectories with a stochastic differential equation

$$dX_t = v(X_t, t)dt + \epsilon dB_t, \tag{7}$$

where X_t denotes a cell state at time t, B_t is Brownian motion, and ϵ is a diffusion coefficient. As t varies, X_t describes a path, or trajectory, through $\mathcal{X}$, and the SDE (7) induces a probability law $\mathbb{P}$ on such trajectories. Our goal will be to recover this law on trajectories from independent samples from the marginals $\mathbb{P}_t$ collected at various time-points, as described in Sect. 2.

In general, the trajectories are not uniquely determined by the marginals. For example, suppose the distribution $\mathbb{P}_t$ is uniform on the unit circle for all t. We do not know if the cells are spiraling clockwise, or counter clockwise, or doing something else. However, if the vector field v for the SDE is *conservative*, meaning that it is the gradient of a potential Φ, then the curve of marginals uniquely determines the trajectories (Theorem 2).

To state this theorem precisely, let $\mathbb{P}$ be the law of the SDE

$$dX_t = -\nabla\Phi(X_t, t) + \epsilon dB_t, \tag{8}$$

and we denote by $\mathbb{P}_t$ the temporal marginal at time t (Figure XXX). We show that $\mathbb{P}$ can be identified by finding the closest process to the Wiener process $\mathbb{W}^\epsilon$ in the sense of relative entropy $H(\cdot|\mathbb{W}^\epsilon)$, which is defined as

$$H(\mathbb{P}|\mathbb{W}^\epsilon) = \int \log(\frac{d\mathbb{P}}{d\mathbb{W}^\epsilon})d\mathbb{P}.$$

Theorem 2 *If $\mathbb{P}$ is the law of the SDE in equation* (8) *with initial condition $H(\mathbb{P}_0|\mathbb{W}_0^\epsilon) < \infty$ and $\mathbb{R}$ is any other law on paths with $\mathbb{R}_t = \mathbb{P}_t$ for all t, then*

$$H(\mathbb{P}|\mathbb{W}^\epsilon) \leq H(\mathbb{R}|\mathbb{W}^\epsilon),$$

with equality if and only if $\mathbb{P} = \mathbb{R}$.

In other words, given knowledge of the marginals $\mathbb{P}_t$, one can reconstruct $\mathbb{P}$ by minimizing the strictly convex functional $H(\cdot|\mathbb{W})$ among all laws having these temporal marginals. For a proof of this theorem, see [7].

Minimizing the relative entropy $H(\mathbb{R}|\mathbb{W})$ can be understood to measure the *length* of the curve $\mathbb{R}$ with entropy regularized optimal transport. For simplicity, we will explain this in the case of coupling discrete measures μ_0 and μ_1 with a single Schrodinger bridge. Ordinary entropy-regularized optimal transport is an optimization over coupling matrices π

$$\begin{aligned} \underset{\pi}{\text{minimize}} \quad & \sum_{x,y} c(x,y)\pi(x,y) + \epsilon H(\pi) \\ \text{s.t.} \quad & \sum_{y} \pi(x,y) = \mu_0(x) \\ & \sum_{x} \pi(x,y) = \mu_1(y) \end{aligned} \tag{9}$$

where $\mathcal{H}(\pi) = \int \pi(x,y)\log \pi(x,y) dx dy$ is the entropy of π.

The objective can be rewritten (adding a constant) to obtain an equivalent optimization problem in terms of minimizing relative entropy between the coupling matrix π and $K = e^{-\frac{C}{\epsilon}}$, where C is the cost matrix.

$$\begin{aligned} \underset{\pi}{\text{minimize}} \quad & \mathrm{H}(\pi|K) \\ \text{s.t.} \quad & \sum_{y} \pi(x,y) = \mu_0(x) \\ & \sum_{x} \pi(x,y) = \mu_1(y) \end{aligned} \tag{10}$$

where $\mathcal{H}(\pi|K) = \sum_{i,j} \pi_{i,j} \log \frac{\pi_{i,j}}{K_{i,j}}$. Note that the optimal $\pi^\star$ from (9) is the same as the optimal $\pi^\star$ from (10).

Finally, the optimization problem can be reformulated as an optimization over laws on paths $\mathbb{R}$

$$\begin{aligned} \underset{\mathbb{R}}{\text{minimize}} \quad & \mathrm{H}(\mathbb{R}|\mathbb{W}^\epsilon) \\ \text{s.t.} \quad & \mathbb{R}_0 = \mu_0 \\ & \mathbb{R}_1 = \mu_1 \end{aligned} \tag{11}$$

Note that the objective here is relative entropy to the Wiener process $\mathbb{W}^\epsilon$ with diffusion coefficient ϵ, which is the same as the regularization parameter in the entropy regularized optimal transport problem (9).

The optimization problem (11) is a reformulation of (9) and (10) in the sense that the optimal law on paths $\mathbb{R}^\star$ can recover the optimal coupling matrix $\pi^\star$ and vice versa. Pick a random path X_t distributed according to $\mathbb{R}^\star$. Then $X_0 \sim \mu_0$ and $X_1 \sim \mu_1$ because $\mathbb{R}^\star$ is feasible for (11). Moreover, the joint distribution of X_0 and X_1 is $\pi^\star$. Conversely, let $X_0 \sim \mu_0$ and $X_1 \sim \mu_1$ be jointly distributed according to π. Connect X_0 and X_1 with a Brownian bridge. The resulting random path is distributed according to $\mathbb{R}$.

Finally, going back to Theorem 2, the curve $\mathbb{R}$ is no longer only constrained at the end-points $\mathbb{R}_0$ and $\mathbb{R}_1$, but instead at all times t we have $\mathbb{R}_t = \mathbb{P}_t$. When we observe the process $\mathbb{P}_t$ at finitely many time-points, we can connect the empirical distributions $\hat{\mathbb{P}}_{t_i}$ with entropy regularized OT as in Fig. 7. This is the curve minimizing relative entropy and passing precisely through all the data $\hat{\mathbb{P}}_{t_i}$. The resulting estimator would be consistent in the limit of infinitely many samples per time-point *and* infinitely many time-points. It turns out, however, that we do not need to send the number of samples per time-point to infinity.

Instead of precisely passing through the data $\hat{\mathbb{P}}_{t_i}$, we can find a curve that passes *close* to the data and still has small length, measured by $H(\mathbb{R}|\mathbb{W}^\epsilon)$. This is especially important in the case of the high-density time-courses of Fig. 14; instead of 'connecting the dots', we define an estimator which trades off data fitting with regularization. It passes close to the data and still minimizes the length of the curve. This estimator can be thought of as performing a regression in the space of developmental curves. The following theorem establishes consistency for this estimator without requiring the number of samples per time-point to increase – only the number of time-points is sent to infinity.

In order to state the theorem precisely, suppose we collect scRNA-seq profiles at T time-points and obtain empirical distributions $\hat{\mathbb{P}}_{t_1}, \ldots, \hat{\mathbb{P}}_{t_T}$. We define the estimator $\mathbb{R}^\star$ to minimize a trade-off between data-fitting and regularization as follows

$$\mathbb{R}^\star = \underset{\mathbb{R}}{\operatorname{argmin}} \quad \mathrm{Fit}_{\lambda,\sigma}(\mathbb{R}_{t_1}) + \epsilon H(\mathbb{R}|\mathbb{W}^\epsilon) \tag{12}$$

where Fit is a data fitting term defined in equation (13) below. The parameter ϵ controls the level of regularization, measured by $H(\mathbb{R}|\mathbb{W}^\epsilon)$.

Roughly speaking, the data fitting is the likelihood. More precisely, let

$$\mathrm{Fit}_{\lambda,\sigma}(\mathbb{R}_{t_1}, \ldots, \mathbb{R}_{t_T}) = \frac{1}{\lambda} \sum_{i=1}^{T} (\Delta t_i) \tilde{\mathrm{Fit}}_\sigma(\mathbb{R}_{t_i} | \mathbb{P}_{t_i}), \tag{13}$$

where $\Delta t_i = t_i - t_{i-1}$ and where

$$\tilde{\mathrm{Fit}}_\sigma(\mathbb{R}_{t_i} | \mathbb{P}_{t_i}) = \int -\log \left(\int \exp\left(-\frac{\|x-y\|^2}{2\sigma^2} \right) d\mathbb{R}_{t_i}(x) \right) d\hat{\mathbb{P}}_{t_i}.$$

This is the negative log-likelihood of observing $\hat{X}_{t_i,j} = X_{t_i,j} + \sigma Z_{i,j}$ with $X_{t_i,j} \sim \mathbb{R}_{t_i}$ and $Z_{i,j} \sim N(0, 1)$.

We are now ready to state the consistency theorem for the estimator $\mathbb{R}^\star$ defined in equation (12).

Theorem 3 *If* $\{t_i\}_{i=1}^T$ *becomes dense in* $[0, 1]$ *as* T *grows, then*

$$\lim_{\lambda,\sigma\to 0} \lim_{T\to\infty} \mathbb{R}^\star = \mathbb{P}.$$

For a proof of this theorem, see [7].

This theorem guarantees consistency for the estimator (12), even if we only sample one cell per time-point. An interesting open question is: what is the optimal number of samples per time-point? Suppose one has a fixed budget N samples to collect over T time-points. How should one choose T depending on N? Based on the theorem above, we conjecture that $T = N$ is optimal. In other words, one should collect as many time-points as possible. This motivates the development of new experimental technologies to collect as many time-points as possible, because it is currently practically very difficult to collect many time-points.

4.1 Algorithms

The optimization problem in equation (12) is convex, but its argument $\mathbb{R}$ is a complex infinite dimensional object — a probability law on paths. One can formulate an equivalent convex problem over a reduced space: through the lense of developmental curves and Theorem 2, one needs only to reconstruct the marginals $\mathbb{P}_t$ in order to reconstruct the trajectories. Indeed, we can set up an optimization over marginals $\mathbb{R}_{t_i}$ at the observed time-points t_i, as illustrated in Fig. 16.

To state this precisely, let $F(\mu)$ be the reduced objective over a vector of marginals $\mu = \{\mu_1, \ldots, \mu_T\}$ defined as follows

$$F(\mu) = \text{Fit}(\mu_1, \ldots, \mu_T) + \sum_{i=1}^{T-1} \frac{1}{\Delta t_i} T_{\epsilon \Delta t_i}(\mu_i, \mu_{i+1}) + \epsilon \sum_{i=1}^{T} H(\mu_i) \tag{14}$$

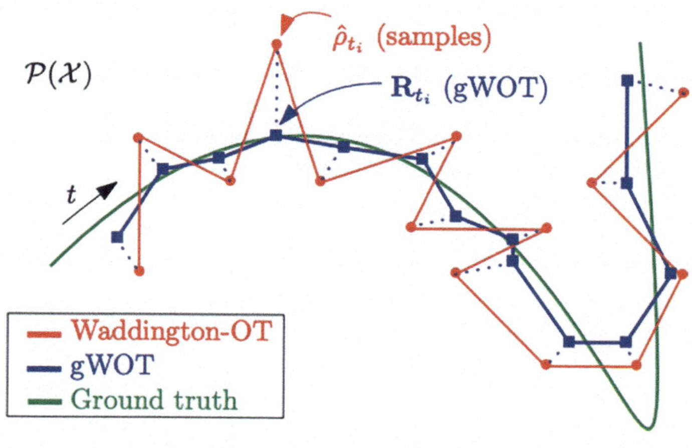

Fig. 16 A developmental curve recovered from data. WaddingtonOT (red) directly connects the empirical distributions, which can result in over-fitting. Global WaddingtonOT (blue) finds the shortest curve passing close to the data

where Fit is defined above in equation (13) and where $T_{\epsilon\Delta t_i}(\mu_i, \mu_{i+1})$ is the entropy regularized cost of transport from μ_i to μ_{i+1} defined as follows

$$T_\tau(\mu_i, \mu_{i+1}) = \min_\pi \int c_\tau(x, y) d\pi(x, y) + \tau H(\pi | \mu_i \otimes \mu_{i+1}).$$

Here $c_\tau \underset{\tau \to 0}{\to} \frac{1}{2}\|y - x\|^2$ is the log-heat-kernel on $\mathcal{X}$. Since Δt_i is typically small, so $\tau = \epsilon \Delta t_i$ is small, we can use $c_\tau(x, y) \approx \frac{1}{2}\|x - y\|^2$ in practice.

Theorem 4 *There is a computable bijection between minimizers of the path-space formulation* (12) *and minimizers of F from equation* (14).

For a proof of this theorem, see [4].

Equipped with this theorem, we see that we can infer trajectories by fitting a piece-wise geodesic developmental curve to the data, as illustrated in Fig. 16. The resulting optimization problem—minimizing $F(\mu)$ over marginals $\mu_1, \ldots, \mu_T$—is convex, and connecting the marginals μ_i (blue squares in Fig. 16) gives the piece-wise geodesic curve.

Solving for the marginals. The practical question, then, is how can we solve for the marginals μ_i in equation (14)? The simplest approach is to parameterize μ_i as a sum of Diracs $\mu_i = \sum_{k=1}^n w_k \delta_{x_k}$. If we fix the support $\{x_1, \ldots, x_n\}$, then the weights w_k can be identified through ordinary finite-dimensional convex optimization. However, identifying the support x_k requires a more sophisticated technique, because the objective $F(\mu)$ is not jointly convex in w_k and x_k.

It turns out that a technique called *mean field Langevin dynamics* makes it possible to solve for w_k and x_k. The goal of Langevin dynamics is: given an objective $V \in \mathcal{C}^2(\mathcal{X}$, sample from $e^{-V/\tau}$ for some temperature parameter $\tau > 0$. Intuitively, Langevin dynamics can be constructed as a limit of noisy gradient steps. Let

$$Y_{k+1} = -\eta \Delta V(Y_k) + \sqrt{2\tau\eta} Z_k, \qquad Y_0 \sim \mu_0, \quad Z_k \sim \mathcal{N}(0, I).$$

As the step size shrinks, $\eta \to 0$, this converges in law to a continuous-time process Y_s called Langevin dynamics. Let $s = k\eta$, then

$$dY_s = -V(Y_s)ds + \sqrt{2\tau} dB_s \qquad Y_0 \sim \mu_0, \quad B_s \text{ Brownian motion.}$$

Moreover, $\mu_s = \text{Law}(Y_s)$ follows the Fokker-Planck equation:

$$\delta_s \mu_s = \nabla \cdot (\mu_s \nabla V) + \tau \Delta \mu_s. \tag{15}$$

Note that one should be careful not to confuse this 'optimization time 's with developmental time t. Similarly, the developmental process X_t, indexed by developmental time, is different from this Langevin process Y_s, indexed by optimization time.

One can interpret μ_s as Wasserstein gradient flow of

$$F_\tau(\mu) = \int V d\mu + \tau H(\mu) = H(\mu|\mu^\star{}_\tau), \tag{16}$$

where $\mu^\star{}_\tau \propto e^{-V/\tau}$, $H(\mu) = \int \log(d\mu/dx)d\mu$ is the neg-entropy, and $H(\mu|\nu) = \int \log(d\mu/d\nu)$ is the relative entropy. So as $s \to \infty$, μ_s goes to a minimizer of F_τ. The following theorem characterizes the minimizer of F_τ and also quantifies the rate of convergence of μ_s to the minimizer.

Theorem 5 *F_τ admits a unique minimizer $\mu^\star{}_\tau \propto e^{-V/\tau}$. Moreover, if $\mu^\star{}_\tau$ satisfies a ρ_τ log-Sobolev inequality, then*

$$F_\tau(\mu_s) - (\mu^\star{}_\tau) \leq e^{-2\tau\rho_\tau s}(F_\tau(\mu_0) - F_\tau(\mu^\star{}_\tau)).$$

This theorem was proven in 1989 by Holley, Kusuoka, and Stroock [6]. Note that the function F_τ defined in equation (16) is the sum of a linear part plus τ times the entropy of μ. Recent work by Chizat generalizes this to any convex function of μ plus τ times entropy of μ [2].

Back to our trajectory inference objective (Equation (14), our objective can be written as $F(\mu) = G(\mu) + \epsilon H(\mu)$, where

$$G(\mu) = \text{Fit}(\mu_1, \ldots, \mu_T) + \sum_{i=1}^{T-1} \frac{1}{\Delta t_i} T_{\epsilon \Delta t_i}(\mu_i, \mu_{i+1}).$$

Note that G is not convex, but $G + \epsilon H$ is convex. Therefore, we can apply MFL to $F_\tau = G + (\tau + \epsilon)H$ for some $\tau > 0$.

Theorem 6 *If $\mathcal{X}$ is compact, the Mean-Field Langevin dynamics for F_τ is well-posed and converges exponentially to global minimizers.*

For a proof of this theorem and further discussion, see [2].

Equipped with this theorem, we have end-to-end theoretical guarantees for trajectory inference: we have an algorithm which provably solves the optimization problem. The solution to the optimization problem recovers the true trajectories.

References

1. R. Bargaje et al., Cell population structure prior to bifurcation predicts efficiency of directed differentiation in human induced pluripotent cells. Proc. Natl. Acad. Sci. **114**(9), 2271–2276 (2017). ISSN: 0027-8424. https://doi.org/10.1073/pnas.1621412114. eprint: https://www.pnas.org/content/114/9/2271.full.pdf. https://www.pnas.org/content/114/9/2271
2. L. Chizat, Mean-field Langevin dynamics: exponential convergence and annealing. Trans. Mach. Learn. Res. **2022** (2022). https://api.semanticscholar.org/CorpusID:246473134

3. L. Chizat et al., Scaling algorithms for unbalanced optimal transport problems. Math. Comput. **87**(314), 2563–2609 (2018)
4. L. Chizat et al., Trajectory inference via mean-field Langevin in path space, in *Advances in Neural Information Processing Systems*, ed. by S. Koyejo et al. vol. 35 (Curran Associates, Inc., 2022), pp. 16731–16742. https://proceedings.neurips.cc/paperË™files/paper/2022/file/6a5181cfe76f67b37a7e1bb19837abdf-Paper-Conference.pdf, 35 (2022)
5. P.B. Gupta et al., Stochastic state transitions give rise to phenotypic equilibrium in populations of cancer cells. Cell **146**(4), 633–644 (2021/05/26 2011). https://doi.org/10.1016/j.cell.2011.07.026
6. R.A. Holley, S. Kusuoka, D.W. Stroock, Asymptotics of the spectral gap with applications to the theory of simulated annealing. J. Funct. Anal. **83**(2), 333–347 (1989). ISSN: 0022-1236. https://doi.org/10.1016/0022-1236(89)90023-2. https://www.sciencedirect.com/science/article/pii/0022123689900232
7. H. Lavenant et al., Towards a mathematical theory of trajectory inference (2021)
8. C. Léonard, A survey of the Schrol̀ˆdinger problem and some of its connections with optimal transport (2013). arXiv:1308.0215
9. E.Z. Macosko et al., Highly parallel genome-wide expression profiling of individual cells using nanoliter droplets. Cell **161**(5), 1202–1214 (2015). ISSN: 0092-8674. https://doi.org/10.1016/j.cell.2015.05.002. http://www.sciencedirect.com/science/article/pii/S0092867415005498
10. A.J. Massri et al., Developmental Single-cell transcriptomics in the Lytechinus variegatus Sea Urchin Embryo. bioRxiv (2020). https://doi.org/10.1101/2020.11.12.380675. eprint: https://www.biorxiv.org/content/early/2020/11/13/2020.11.12.380675.full.pdf. https://www.biorxiv.org/content/early/2020/11/13/2020.11.12.380675
11. C. Qiu et al., A single-cell time-lapse of mouse prenatal development from gastrula to birth. Nature **626**(8001), 1084–1093 (2024)
12. A.B. Rosenberg et al., Single-cell profiling of the developing mouse brain and spinal cord with split-pool barcoding. Science **360**(6385), 176–182 (2018)
13. G. Schiebinger, Reconstructing developmental landscapes and trajectories from single-cell data. Curr. Opin. Syst. Biol. 27, 100351 (2021). ISSN: 2452-3100. https://doi.org/10.1016/j.coisb.2021.06.002. https://www.sciencedirect.com/science/article/pii/S2452310021000457
14. G. Schiebinger, J. Gould, *Waddington OT Tutorial* (2019). https://broadinstitute.github.io/wot/tutorial/. Accessed 19 June 2024
15. G. Schiebinger et al., Optimal-transport analysis of single-cell gene expression identifies developmental trajectories in reprogramming. Cell **176**(4), 928–943 (2019)
16. K. Takahashi et al., Induction of pluripotent stem cells from adult human fibroblasts by defined factors. Cell **131**(5), 861–872 (2007)
17. Y. Tang et al., Potential landscape of high dimensional nonlinear stochastic dynamics with large noise. Sci. Rep. **7**(1), 15762 (2017)
18. C.H. Waddington, *The Strategy of the Genes* (George Allen & Unwin, 1957)
19. J. Zhou et al., Systematic drug perturbations on cancer cells reveal diverse exit paths from proliferative state. Oncotarget **7** (2016). https://doi.org/10.18632/oncotarget.7294

The manufacturer's authorised representative in the EU is Springer
Nature Customer Service Centre GmbH, Europaplatz 3, 69115 Heidelberg,
Germany. If you have any concerns regarding our products, please
contact ProductSafety@springernature.com

Printed and bound by CPI Group (UK) Ltd, Croydon, CR0 4YY
17/09/2026
02218460-0001